Springer Series in Advanced Manufacturing

Other titles in this series

Massimiliano Caramia and Paolo Dell'Olmo

Effective Resource Management in Manufacturing Systems

Optimization Algorithms for Production Planning

With 78 Figures

 Springer

Massimiliano Caramia, PhD
Istituto per le Applicazioni del Calcolo
"M. Picone" – CNR
Viale del Policlinico, 137
00161 Roma
Italy

Paolo Dell'Olmo, Prof.
Universitá di Roma "La Sapienza"
Dipartimento di Statistica
Probabilitá e Statistiche Applicate
Piazzale Aldo Moro, 5
00185 Roma
Italy

Series Editor:
Professor D. T. Pham
Intelligent Systems Laboratory
WDA Centre of Enterprise in
Manufacturing Engineering
University of Wales Cardiff
PO Box 688
Newport Road
Cardiff
CF2 3ET
UK

British Library Cataloguing in Publication Data
Caramia, Massimiliano
    Effective resource management in manufacturing systems:
    optimization algorithms for production planning. -
    (Springer series in advanced manufacturing)
    1. Production management - Mathematical models
    2. Mathematical optimization 3. Algorithms
    I. Title II. Dell'Olmo, Paolo, 1958-
    658.5'0015181
ISBN-10: 1846280052

Library of Congress Control Number: 2005933474

Springer Series in Advanced Manufacturing ISSN 1860-5168
ISBN-10: 1-84628-005-2   e-ISBN 1-84628-227-6   Printed on acid-free paper
ISBN-13: 978-1-84628-005-4

Printed in Germany

9 8 7 6 5 4 3 2 1

Springer Science+Business Media
springeronline.com

To our wives and sons

Preface

Manufacturing systems, regardless of their size, have to be able to function in dynamic environments with scarce resources, and managers are asked to assign production facilities to parallel activities over time respecting operational constraints and deadlines while keeping resource costs as low as possible.

Thus, classic scheduling approaches are not adequate when (i) a task simultaneously requires a set of different resources and (ii) a trade-off between different objectives (such as time, cost and workload balance) has to be made. In such cases, more sophisticated models and algorithms should be brought to the attention of the managers and executives of manufacturing companies.

In this framework, this book aims to provide robust methods to achieve effective resource allocation and solve related problems that appear daily, often generating cost overruns. More specifically, we focus on problems like on line workload balancing, resource levelling, the sizing of machine and production layouts, and cost optimization in production planning and scheduling. Our approach is based on providing quantitative methods, covering both mathematical programming and algorithms, leading to high quality solutions for the problems analyzed. We provide extensive experimental results for the proposed techniques and put them in practical contexts, so that, on the one hand, the reader may reproduce them, and, on the other hand, the reader can see how they can be implemented in real scenarios.

In writing this book, an attempt has been made to make the book self contained, introducing the reader to the new modelling approaches in manufacturing, presenting updated surveys on the existing literature, and trying to describe in detail the solution procedures including several examples. Yet, the complexity of the topics covered requires a certain amount of knowledge regarding several aspects which, in this book, are only partially covered. In particular, for general quantitative modelling approaches in the management of manufacturing systems and for the large body of machine scheduling models and algorithms the reader is encouraged to refer to books and articles cited in the text.

The book can be used by Master and PhD students in fields such as Manufacturing, Quantitative Management, Optimization, Operations Research, Control Theory and Computer Science, desiring to acquire knowledge in updated modelling and optimization techniques for non classical scheduling models in production systems. Also, it can be effectively adopted by practitioners having responsibility in the area of resource management and scheduling in manufacturing systems. To facilitate both thorough comprehension and the application of the proposed solutions into practice, the book also contains the algorithms source code (in the C language) and the complete description of the mathematical models (in the AMPL language). It should also be noted that most of the selected topics represent quite general problems arising in practice. Application examples have been provided to help the mapping of real world problem recognition and modelling in the proposed framework.

As with any book, this one reflects the attitudes of the authors. As engineers, we often made the underlying assumption that the reader shares with us a "system view" of the world, and this may not always be completely true. As researchers in discrete optimization, we place strong emphasis on algorithm design, analysis and experimental evaluation. In this case, we made the assumption that someone else took charge of providing the data required by the algorithms. Although this is a preliminary task that a manager cannot underestimate, we believe that current production systems are already equipped with computing and communication capabilities that can make this job manageable.

A glance at the table of contents will provide an immediate list of the topics to be discussed, which are sketched in the following.

In Chapter 1 we describe manufacturing systems according to their general structure, the goals typically pursued in this context, and focus our attention on resource allocation. A brief analysis on algorithmic techniques used in other chapters is presented.

In Chapter 2 we analyze the problem of balancing the load of n machines (plants) in on-line scenarios. We describe known techniques and propose a new algorithm inspired by a metaheuristic approach. The novelty of the approach also stems from the fact that the algorithm is still an on-line constructive algorithm, and thus guarantees reduced computing times, but acts as a more sophisticated approach, where a neighborhood search has to be made in the same way as with a local search method.

Chapter 3 is about resource levelling, i.e., the problem of smoothing the shape of the resource profile in a schedule. This problem is discussed in the scenario in which tasks require more than one resource at a time, and the total amount of resources is limited. The problem is off-line. We propose a metaheuristic approach to the problem and a comparison with the state of the art.

Chapter 4 studies the problem of scheduling jobs in a robotized cell with m machines. The problem consists of a case where each part entering the production system must be loaded by means of a robot on one available machine, and, when the machine finishes the execution of its task, the part must be unloaded and must exit the system.

The last part of the book is dedicated to tool management on flexible machines (Chapter 5). We study the problem of managing tool changeovers and consequently setup times, in flexible environments, where parts are not produced in batch. Different heuristics are proposed for this problem, and a comparison with known algorithms for the same problem is presented.

Rome,   *Massimiliano Caramia*
May 2005   *Paolo Dell'Olmo*

Contents

List of Figures

List of Tables

1

Manufacturing Systems: Trends, Classification, and Behavior Patterns

World-wide competition among enterprises has led to the need for new systems capable of performing the control and supervision of manufacturing processes through the integration of information and automation islands.

Increasing market demands must be met by manufacturing enterprises to avoid the risk of becoming less competitive.

The adoption of new manufacturing concepts combined with the implementation of emergent technologies is the answer to the need for an improvement in productivity and quality and to a corresponding decrease in price and delivery time.

Nowadays, the *distributed manufacturing* organisational concept, *e.g.* the *Virtual Enterprise* (see Figure 1.1), requires the development of decentralised control architectures, capable of reacting to disturbances and changes in their environment, and capable of improving their performance.

Mass manufacturing, idealised by Henry Ford, was a strap down system, incapable of dealing with variations in the type of product. This rigidity started becoming an obstacle and with onset of worldwide competitiveness mass manufacturing became viable for only some products. This is how mass manufacturing evolved from the era of manufacturing to the era of mass customisation.

Nowadays, each product comes out in several models, and each model can be highly customised in order to satisfy the requirements of customers; a good example being the automobile industry. This requires the implementation of an automated job shop type of production in a truly coordinated production system.

Coordination and effectiveness can only be achieved through the appropriate organization and management of a manufacturing system, that is, through the efficient and effective implementation of manufacturing control.

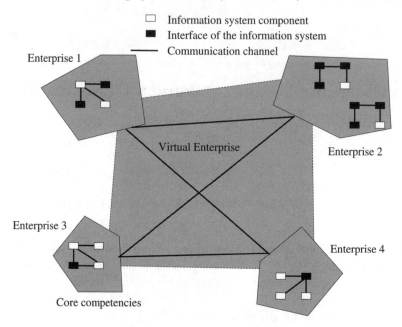

Fig. 1.1. Example of a Virtual Enterprise. The Virtual Enterprise is a temporary network of independent companies linked by information technology to share skill, costs and access to one another markets

1.1 Distributed Flexible Manufacturing Systems

In today's global and highly competitive market, enterprises must be aware of rapidly evolving market opportunities, and react quickly and properly to customers' demands.

A *Distributed Flexible Manufacturing System* (DFMS) is a goal-driven and data-directed dynamic system which is designed to provide an effective operation sequence to ensure that products fulfil production goals, meet real-time requirements, and allocate resources optimally [24].

In contrast to a centralized or hierarchical structure, the distributed control architecture reveals its advantages not only in providing efficient and parallel processing, but also in achieving flexible system integration maintenance, thus providing the right solutions to the requirements of competitiveness.

Indeed:

- An increase in product diversity over time expands the associated risks and costs, which are sometimes prohibitive, while distributing responsibilities over multiple entities, allowing the risks and costs to become acceptable and opening up market opportunities.

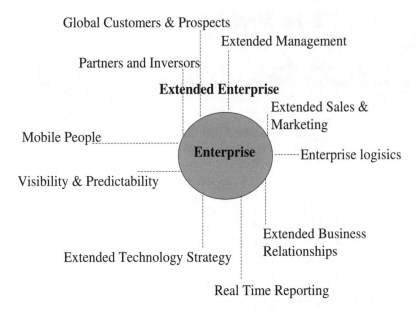

Fig. 1.2. The Extended Enterprise

- An increase in technological complexity forces enterprises to acquire knowledge in non-fundamental domains, which implies increased time-to-market periods, while distributing competencies over different enterprises, allows each one to maintain its core competency while taking advantage of market opportunities.
- Market globalisation virtually increases both enterprise opportunities and risks. Each enterprise has to operate in the global market with globally based enterprises supplying global products. However, developing relationship and partnerships with such enterprises sharing challenges and risks while allowing them to benefit from a wider market.

Different management approaches have been adopted, depending on the different levels of partnership, trust and dependency between enterprises:

- *Supply Chain* management, characterized by a rudimentary relationship between the supplied entity and the supplier, task and technological competency distribution, and centralized strategies and risks (see Figure 1.3).
- The *Extended Enterprise*, where entities develop more durable, coupled and mutual intervening relationships, sharing technological and strategical efforts. However, the supplied entity maintains a dominant position over suppliers (see Figure 1.2).

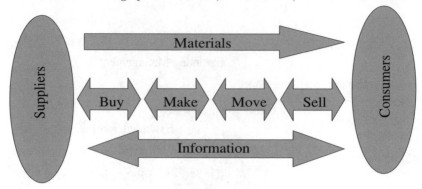

Fig. 1.3. The Supply Chain

- The *Virtual Enterprise*, a very dynamic and restructuring organization, where the supplier and supplied entity are undifferentiated and no dominant position exists (see Figure 1.1).

Although the previous description relates to an inter-enterprise context, the same characteristics and behaviour patterns (distribution, decentralization, autonomy and dependency) are also apparent in an intra-enterprise context. Intra-enterprise workgroups emphasize self-competencies while combining their efforts towards a global response to external requirements.

The *Distributed Flexible Manufacturing System* is an abstract concept (*i.e.* a class of systems) characterized by a set of common features and behaviour patterns, with several specific features (*i.e.* instantiations), named *organizational paradigms*.

1.1.1 DFMS Properties

DFMS are characterised by several properties and behaviour patterns. Such features relate to both the overall system and each composing entity. We may classify those properties as *basic properties*, *i.e.*, those properties that define a DFMS, and as *characterizing properties*, *i.e.*, those properties that distinguish DFMSs from each other (see Table 1.1).

Basic properties

Autonomy – An entity is said to be autonomous if it has the ability to operate independently from the rest of the system and if it possesses some kind of control over its actions and internal state [50], *i.e.*, autonomy is the ability of an entity to create and control the execution of its own plans and/or strategies, instead of being commanded by another entity

Table 1.1. Basic and characterizing properties of DFMS

Basic properties	Characterizing properties
Antonomy	Flexibility
Distibution	Adaptability
Decentralization	Agility
Dynamism	
Reaction	

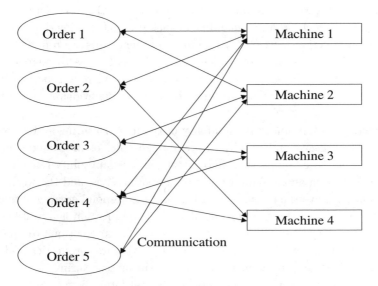

Fig. 1.4. Configuration of an autonomous distributed manufacturing system

(*e.g.*, a master/slave relationship). In Figure 1.4 we depict an example of configuration of an autonomous distributed manufacturing system.

Distribution – A system is said to be distributed if different entities operate within the system;

Decentralization – Decentralization means that an operation/competency can be carried out by multiple entities.

Dynamism – Refers to changes in the structure and behaviour of a manufacturing system during operation. This expresses different competencies, responsibilities and relationships between entities.

Reaction – An entity is said to be reactive if it adjusts its plans in response to its perceptions.

Characterizing properties

Flexibility – Flexibility is the ability the system exhibits during operation to change processes easily and rapidly under a predefined set of possibilities. Each one is specified as a routine procedure, and defined ahead of time so that the need to manage it are in place. In manufacturing, Flexibility is related to physical flexible machinery. Flexible means that machines (or cells) are able to execute several different operations. In addition, they can quickly change from one production plan to another depending on the type of part that has to be manufactured at a given point in time. The concept of the Flexible Manufacturing System (FMS) is very popular with companies which produce small lots of each product, mixing different lots in the production flow. One of the main problems in achieving flexibility is related to transportation. Since a product has to pass through several workstations in order to be manufactured and different products have different routes, the transport links between workstations should be as *free* as possible.

Adaptability – Adaptability is the ability of the manufacturing system to be maintained easily and rapidly in order to respond to manufacturing requirements based on its shop floor constraints. Adaptability refers to the reconfiguration and scalability of production facilities, and the workforce that has to have the incentive and flexibility to respond creatively to customer needs and thus requires flexibility. Ill-specification is a well-known synonym. A system is said to be adaptable if it can continue to operate in the face of disturbances and change its structure, properties and behaviour according to the new situations it encounters during its *life-time*. A disturbance is any event not previously and formerly specified (*e.g.*, a machine breakdown or a new type of product to manufacture). However, it is very hard to predict every disturbance that may occur.

Agility – Agility is understood as a management paradigm consisting of different approaches in multiple organisational domains. However, Agility presumes system empowerment, achieved through continuous observation in search of new market opportunities. Agile manufacturing enterprises continually evolve to adapt themselves and to pursue strategic partnerships in order to prosper in an extremely dynamic and demanding economy.

Flexibility is the simplest approach and relates directly to the shop floor. It allows the shop floor to react according to a predefined set of possibilities to deal with primary disturbances in production. Feedback from the shop floor comes mainly from the identification of the current lot, which serves as a basis for the decision to download the correct production plan.

On the contrary, Adaptability is based on sub-specification, *i.e.*, the system is not completely defined, which allows run-time specification and change according to real time requirements. Adaptability means being able to understand disturbances on the shop floor, and generating new production plans for the current situation, if necessary.

Agility is the uppermost concept and relates to strategic options. Perceiving its business environment, the enterprise has to continuously evolve, adapting internally and pursuing external strategic partnerships that complete its own competencies.

Adaptability also plays an important role by understanding the business strategies generated by the Agility. Using these strategies, new production plans or modified ones are added to the production plan database to be used on the shop floor. These new plans may be alternative ways of making current products, or plans for the production of new products.

1.1.2 DFMS Behaviour

A Flexible Distributed Manufacturing System may exhibit several behaviour patterns, the most important, for the scope of our study, being cooperation and coordination. However, the goal of any system (not only in manufacturing) is to behave coherently. *Coherence* refers to how well a set of entities behaves as a whole [162]. A coherent system will minimise or avoid conflicting and redundant efforts among entities [140].

A coherent state is achieved by engaging in one or more of the following behaviour patterns:

Cooperation – A process whereby a set of entities develops mutually acceptable plans and executes these plans [169]. These entities explicitly agree to try to achieve a goal (or several goals) with the partial contribution of each participant. The goal need not be the same for each participant, but every participant expects to benefit from the cooperation process.

Coordination – Is the process involving the efficient management of the interdependencies between activities [128].

Competition – Is a process whereby several entities independently try to achieve the same goal (with or without the knowledge of the other participants - *explicit* or *implicit* competition).

During the natural execution of the system, the majority of these behaviour patterns are observed. For instance, in order to compare a DFMS with other distributed systems, consider a distributed problem-solver [93]: it will exhibit both cooperation and coordination. In fact, the several solvers cooperate by sharing their efforts and also by coordinating their activity dependencies, *e.g.* on a finite element analysis [103], each solver operates only on a small set of data, exchanging some data with its neighbours. They co-ordinate their activities so that each one's output is valid and cooperate by dividing the workload.

Cooperation and Coordination in manufacturing are intuitively proven necessary when adopting a distributed solution [160]: the resources must *co-operate* to manufacture a product and must *coordinate* their actions since the dependencies between operations must be observed; looking at the multi-enterprise level, each company involved establishes plans accepted by the other partners for the fulfilment of the global objectives, thus they *cooperate* to manufacture a product (*e.g.*, providing different assembled parts) and/or to provide some kind of necessary service (*e.g.*, accounting, distribution, quality testing), and must *coordinate* their activities to avoid inefficiency.

1.1.3 Organizational Paradigms

Distributed Manufacturing has been adopted for a long time, but its organizational structures were only formalized and proposed as an essential solution in recent years.

In order to achieve the above-mentioned characteristics and behaviour patterns, several organizational paradigms have been proposed, namely the Fractal Factory [158, 173], Bionic Manufacturing Systems [141] and Holonic Manufacturing Systems [53, 154], which are included in a broader term designated *Open Hierarchical Systems*.

These paradigms can be distinguished from each other according to their source of origin, *i.e.* mathematics for the fractal factory, nature for the bionic and genetic production systems, and the philosophical theory on the creation and evolution of complex adaptive systems for holonic manufacturing.

The *Fractal Factory* is an open system which consists of independent self-similar units, the fractals, and is a vital organism due to its dynamic organisational structure. Fractal manufacturing uses the ideas of mathematical chaos: the companies could be composed of small components or fractal objects, which have the capability of reacting and adapting quickly to new environment changes. A fractal object has the following features:

- *self-organised*, which means that it does not need external intervention to reorganise itself.
- *self-similar*, which means that one object in a fractal company is similar to another object. In other words, self-similar means that each object contains a set of similar components and shares a set of objectives and visions.
- *self-optimised*, which means it continuously improves its performance.

The explosion of fractal objects into other fractal objects has the particularity of generating objects which possess an organizational structure and objectives similar to the original ones. For Warneke [173], the factory of the future will present different dynamic organisational structures, adopting the project orientation organisation instead of the traditional function oriented organisation. This approach implies that the organisational structure will encapsulate the process and the technology, therefore forming a cybernetic structure.

Bionic Manufacturing Systems (BMS) indexbionic manufacturing have developed using biological ideas and concepts, and assume that manufacturing companies can be built as open, autonomous, cooperative and adaptative entities, which can evolve. BMSs assign to manufacturing systems the structure and organisational behaviour of living beings, defining a parallelism between biological systems and manufacturing systems. The cell, organ or living being is modelled in BMSs by the *modelon* concept, which is composed of other modelons, forming a hierarchical structure. Each modelon has a set of static properties and behaviour patterns, which can be combined with others, forming distinct entities, also designated as modelons. The notion of DNA inheritance is assigned to the manufacturing context in that the properties and behaviour patterns are intrinsically passed on to developed modelons [166]. The biological concept of *enzymes* and their role in living beings is modelled in manufacturing systems by entities called supervisors, which are responsible for the regulation and control of the system. Furthermore, the supervisors also play an organisational and structural role in the cooperation process within the BMSs, influencing the relationships between modelons and imposing self-division or aggregation in order to adapt and react to the requirements imposed by the environment [161].

The *Holonic Manufacturing System* translates the concepts that Koestler [113] developed for living organisms and social organisations into a set of appropriate concepts for manufacturing industries. Koestler used the word *holon* to describe a basic unit of organisation in living organisms and social organisations, based on Herbert Simon's theories. Simon observed that complex systems are hierarchical systems composed of intermediate stable forms which do not exist as auto-sufficient and non-interactive elements, but are simultaneously a part and a whole. The word holon is the representation of this hybrid nature, as it is a combination of the Greek word "holos", which means whole, and the suffix "on", which means particle. A holon is an autonomous and cooperative entity in a manufacturing system, which includes operational features, skills and knowledge, and individual goals. It can represent a physical or logical activity, such as a robot, a machine, an order, a Flexible Manufacturing System, or even a human operator. The holon has information about itself and the environment, containing an information processing part and often a physical processing part. An important feature of an

HMS is that a holon can be part of another holon, *e.g.*, a holon can be broken into several other holons, which in turn can be broken into further holons, which makes it possible to reduce the complexity of the problem.

Note that these paradigms suggest that manufacturing systems will continue to need a hierarchical structure together with the increased autonomy assigned to individual entities. They also suggest that the hierarchy is needed to guarantee the resolution of inter-entity conflicts, and maintain the overall system coherence and objectivity resulting from the individual and autonomous attitude of the entities.

1.1.4 Example of the Implementation of a Holonic Manufacturing System in Induction Motor Production

In the following, we describe a real life problem arising from the production of induction motors to show how a holonic manufacturing system can be implemented.

Induction motors exemplify a low-volume, high-variety production system and highlight many problems that arise from the inflexibility of centralized management system architectures. The system covers the parts production and assembly processes. The input of raw materials includes the bearing bracket (BB), the main bearing (BRG), the frame, copper wire, steel sheets, and steel rods. The production system is made up of the following machine types: casting machines, boring machines, machining centers, fraise machines, clank presses, lathe machines, and grinding machines. The machining processes are needed to create the following parts: bearing brackets (BB), bearings (BRG), stator frames (SS), stator cores, slot insulators, shafts (SFT), rotor cores, rotor conductors, fan covers, and fans, which are then assembled to generate the final product.

In general, a motor plant generates 800 different types of motors, with an order mix of 200 different types of motors per day and 1,000 motor components per day. Lot sizes vary from 1 to 100, however, the average lot size is 5.

The Testbed

As a first step, we build an initial testbed that focuses on the shaft manufacturing section. There are three products that are generated from this testbed:

- P-D24 corresponds to an induction motor shaft with a 24 mm diameter and a length of 327 mm. This shaft is for 1.5 kw induction motors.
- P-SD42 corresponds to an induction motor shaft with 42 mm diameter and a length of 603 mm. This shaft is for 11 kw induction motors.
- P-SD48 corresponds to an induction motor shaft with 48 mm diameter and a length of 676 mm. This shaft is for 22 kw induction motors.

The five machine-types:

- M-1 is a turning center machine.
- M-2 is a grinding machine.
- M-3 is an NC lathe machine.
- M-4 is an NC fraise machine.
- M-5 is a machining center machine.

They are needed for the manufacturing process, and the number of each machine type in the testbed can be specified by the user. There are three alternate production methods shown in Figure 1.5. In the first production method, the raw material, the steel rod, goes to machine type M-1, and is then transferred to machine type M-2. The second production method starts with machine type M-3, then transfers to machine type M-4, and finishes at machine type M-2. The third method also starts at machine type M-3, then goes to machine type M-5, and finishes at machine type M-2.

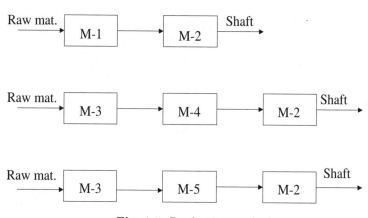

Fig. 1.5. Production methods

System Architecture

A heterarchical architecture was selected for our testbed. It completely eliminates the master/slave relationship between the entities comprising the control system. The result is a flat architecture where modules cooperate as equals, rather than being assigned subordinate and supervisory relationships. Communication and control among modules, to resolve production tasks, is achieved via message transfer schemes. This provides the necessary flexibility and robustness. When a module fails, another module can take over its tasks. In spite of this flexibility, careless implementation of the heterarchical control may produce some sort of cooperative anarchy [67]. Systematic implementation provides several advantages over traditional hierarchical control [23, 65, 66, 67]:

- a higher degree of distribution, modularity, and maintainability,
- a higher robustness and fault tolerance, and
- an implicit modifiability and reconfigurability.

Holons Definition

There are five basic holons defined for this testbed (see Figure 1.6):

- Product Holon, which corresponds to products to be manufactured. For this testbed, they are P-SD24, P-SD42, and P-SD48.
- Machine Holon, which corresponds to machines in the testbed. The five machine types are M-1, M-2, M-3, M-4, and M-5.
- Scheduler Holon, which performs the decision making, product reservation, and resource allocation functions.
- Computing Holon, which calculates the costs and time for a machine to perform a particular machine or assembly tasks.
- Negotiation Holon, which handles negotiation processes between parts and machine holons.

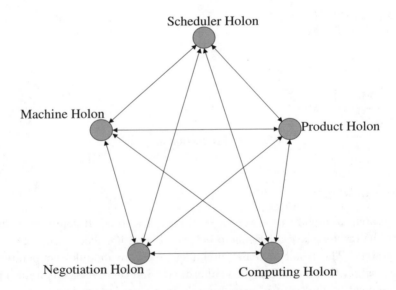

Fig. 1.6. Basic holons

This example shows how the implementation of production systems in a distributed way, *e.g.*, by means of holons, allows one to manage the different functionalites from a twofold viewpoint, *i.e.*, one in which holons act as stand alone entities, and one in which they closely interact.

1.1.5 A Layered Approach to DFMS Modeling

In order to unifying the organisational paradigm concepts presented in Section 1.1.3, and to study the management issues related to a generalized DFMS, we adapt the layered approach presented in [118].

According to this modeling approach, a DFMS can be seen as an aggregation of four layers, such that each element which belongs to a layer at level i represents a layer at level $i + 1$, and is organized as follows.

In the first layer, called the *Multi-Enterprise layer*, the interaction between distributed enterprises is modelled. The enterprises act together in order to achieve a common objective, and each enterprise interacts with its suppliers and customers.

A similar environment is found within each manufacturing enterprise. In fact, each enterprise (element) of the first layer is itself a layer at the second level, called the *Enterprise layer*. In this layer, the cooperation between geographically distributed entities (normally, the sales offices and/or the production sites) takes place.

Zooming into a production site element shows the *Shop Floor layer*, where the distributed manufacturing control (MC) within a production site or shop floor can be found. In this layer, the entities are distributed work areas working together and in cooperation, in order to fulfil all the orders allocated to the shop floor, respecting the due dates.

A single work area contains a *Cell layer*, where the cooperation between equipment, machines and humans takes place.

Figure 1.7 illustrates a shop production floor.

Collapsing the nodes on the top layer onto the elements of the Cell layer, it is easy to see that this approach models a DFMS as a graph where the vertices represent the resources in the system, and links between vertices represent the interaction among the resources. Such a graph is called a *resource graph*, and its structure will be discussed in the following chapters.

Focusing on a 3-layered DFMS allows one to study managerial problems that arise in the distributed manufacturing control function. Therefore, the resource graph will contain *e.g.* as many vertices as the number of interacting facilities in the system, multiplied by the work areas in each facility, multiplied by the robotic cells in each work area.

Fig. 1.7. A shop production floor

1.2 Manufacturing Control (MC)

1.2.1 Definition of Manufacturing Control

In a factory, a lot of activities have to be performed to produce the right goods for the right customer in the right quantities at the right time [13, 45]. Some activities focus on the physical production and manipulation of the products. Other activities refer to the management and control of the physical activities (in Figure 1.10 we show a configuration of a decentralized control system). These production management and control activities can be classified as strategic, tactical and operational activities, depending on the long term, medium term or short term nature of their task [32, 52, 150].

Strategic production management issues relate to the determination of which products should be designed, manufactured and sold, considering the markets and customer expectations. It also relates to the design of the appropriate manufacturing system to produce these products, including the generation of a master schedule to check whether the manufacturing system has enough capacity to meet the estimated market demand.

Tactical production management issues refer to the generation of detailed plans to meet the demands imposed by the master schedule, including the calculation of appropriate release and due dates for assemblies, sub-assemblies and components. These detailed plans may refer to predicted orders, in a make-to-stock environment, or to real customer orders, in a make-to-order environment. Typically, MRP or MRP-II systems (Material Requirements Planning and Manufacturing Resource Planning) carry out these tasks.

Operational production management issues relate to the quasi-real time management of the manufacturing system on the shop floor. It involves commanding the appropriate machines, workers and other resources in a coordinated way, thereby selecting the appropriate resources for each task and the appropriate sequencing of these tasks on each resource. It includes the detailed planning and optimisation needed to fulfil the requirements of the MRP schedule in the best possible way.

Manufacturing Control (MC) refers to the management and control decisions at this operational level, and is the decision making activity concerned with the short-term and detailed assignment of operations to production resources [51].

In Figures 1.8 and 1.9 we compare the functionalities of a conventional manufacturing control system and a holonic manufacturing control system. Manufacturing control functionalities is the topic of the next section.

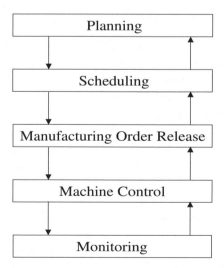

Fig. 1.8. Conventional manufacturing control system

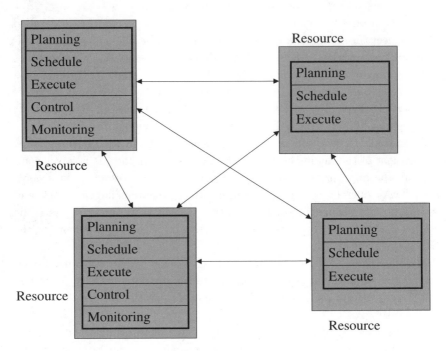

Fig. 1.9. Holonic manufacturing control system

1.2.2 Manufacturing Control Functions

The main function of Manufacturing Control is **resource allocation**. As stated above, resource allocation is also performed at the tactical and strategic level of production management. MC refers to the operational level of resource allocation only. This usually includes a detailed short-term **scheduler**, which plans and optimises the resource allocation beforehand, considering all production requirements and constraints. Typically, scheduling has a total time span ranging from a few weeks to a day, and a time granularity ranging from days to less than a minute. When the schedule is calculated, it is implemented on the shop floor by a **dispatcher**, taking into account the current status of the production system (see Figure 1.11). This function is called **on-line manufacturing control** (OMC). During OMC, the status of the vital components of the system is monitored.

Manufacturing Control also covers process management [132]. Process management includes machine and process monitoring and control, but also the operational level activities of process planning. The **process planning** function defines how products can be made, including the overall process plan

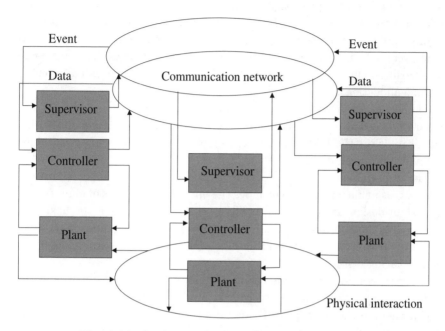

Fig. 1.10. Configuration of a decentralized control system

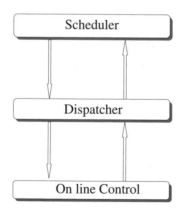

Fig. 1.11. Manufacturing control functions

defining the possible sequences of operations needed to manufacture a product and detailed operation process planning, prescribing the parameters of the individual manufacturing steps. In traditional manufacturing systems, process planning provides manufacturing control with a single sequence of operations to be executed for each product. Recently, process planning has considered more flexible solutions based on the use of alternative routings and alternative machine allocations.

The use of alternative routings, modelled *e.g.* by precedence graphs [71], dynamic precedence graphs [168] or Petri nets [63] enables manufacturing control to select the best sequence of operations depending on the machine load and unforeseen disturbances.

The use of alternative resources for an operation gives similar opportunities. Process management also includes the operational aspects of manufacturing system design, layouting and configuration. Depending on the flexibility of the manufacturing system, the manufacturing system may be adaptable in the short term. These short-term modifications and set-ups are managed by manufacturing control.

MC further needs to provide people and machines with information at the right moment. It therefore has to collect, store and retrieve these data and provide them in the right quantity and at the right level of aggregation. This function includes data capturing from the machines and the analysis of these raw data. It also deals with the management of the programs for NC-machines and with quality and labour management.

Manufacturing control relates to several other enterprise activities, such as marketing, design, process planning, sales, purchasing, medium and long term planning, distribution, and servicing, with which it has to interface. Via the sales function, new orders are introduced into the shop. Orders maybe directly assigned to customers or they may be stock replenishment orders, in a make-to-stock environment. In industry, a combination of a make-to stock and make-to-order environment is very often found and dealt with pragmatically. Traditionally, manufacturing control considers sales as a black box, generating orders with a stochastic distribution. Medium-range planning provides MC control with due dates and release dates for orders. To a large degree, it defines the performance objectives for manufacturing control.

1.2.3 Classification of Production Scheduling

The characteristics of a manufacturing system, such as production volume, flexibility and layout, have a major influence on its control.

The complexity of the resource allocation optimisation (*scheduling*, see below) and on-line control problems depend heavily on the layout. Therefore, manufacturing control strategies differ considerably according to the corresponding layout.

Manufacturing systems and subsystems are often categorised according to their production volume and their flexibility.

Traditionally, the targeted production volume and flexibility of a manufacturing system are inversely proportional: high volume manufacturing systems have limited flexibility.

Typical production types range from continuous production, mass production, and large batch manufacturing, to discrete manufacturing, one-of-a-kind and project-wise manufacturing. This book is restricted to discrete manufacturing.

Within discrete manufacturing, manufacturing systems are classified by their layout (usually related to the way products flow through the manufacturing system).

The rest of this subsection surveys the main production layouts relevant for this book.

The **single machine model** [13] has the most simple layout. It consists of a single machine that performs all operations. Therefore, orders are usually not further decomposed into operations.

The **identical processors model** [13] is similar to the single machine model, but consists of multiple, identical machines.

A **flow shop** consists of different machines processing orders with multiple operations, where:

- each operation has to be executed on a specific machine;
- all the operations of an order have to be executed in a fixed sequence;
- all the orders have to visit the machines in the same sequence.

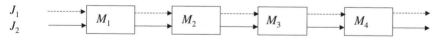

Fig. 1.12. Example of a flow shop with four machines and two jobs

Figure 1.12 depicts a flow shop, whose schedule is reported in Figure 1.14. Figure 1.13 shows an assembly line, a particular type of flow shop.

A **job shop** is the same as a flow shop, except that each order can visit the machines in a different sequence. An example of a job shop is reported in Figure 1.15 and the corresponding schedule in Figure 1.16.

Fig. 1.13. An assembly line

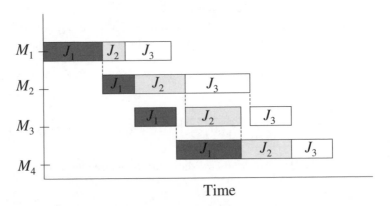

Time

Fig. 1.14. Example of a schedule in a flow shop

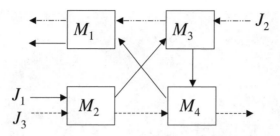

Fig. 1.15. Example of a job shop with four machines and three jobs

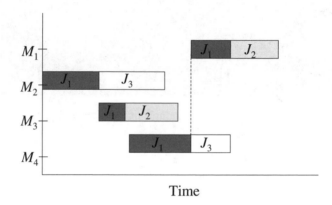

Time

Fig. 1.16. Example of a schedule in a job shop

A **fixed-position shop** is a production systems where processes are executed around the product which does not move in the shop. An example is given by aircraft production (see Figure 1.17).

This book focuses on the job shop layout, as it is the most general case, and is the most suitable system to achieve flexibility in a distributive environment.

These layout models refer to more specific manufacturing systems, like factories with a bottleneck, lines, flexible manufacturing systems (FMSs), or flexible flow shops (FFSs). For some aspects of the control of factories with a bottleneck, the single machine model performs quite adequately [13, 77].

Assembly lines are usually flow shops with limited or no opportunities for orders overtaking each other.

Flexible manufacturing systems are extended to job shops [68], where:

Fig. 1.17. Example of a fixed position shop

- operations of orders follow a sequence allowed by a precedence graph instead of a fixed operation sequence;
- different workstations can perform the same operation;
- transport and set-up have to be modelled.

Recently [117], flexible flow shops have become more important: they inherit the layout and properties of flow shops, but have increased flexibility with respect to machine selection, operation sequencing and sometimes in routing. In this respect, they are mathematically more similar to job shops . However, while all routings are feasible, some routings are highly preferable. In this way, they combine the efficiency and transparency of a flow line with the reactivity and flexibility of a job shop . Such manufacturing systems promise to eliminate the need for a compromise between flexibility and high volume production as is found in traditional manufacturing systems.

For a deeper comprehension of FFSs, the reader can see the case study in Chapter 3.

1.3 Scheduling and Resource Allocation

1.3.1 Definition

Scheduling is defined as the process of *optimising* resource allocation over time beforehand [37, 43, 145].
Resource allocation is deciding when and with which resources all tasks should take place [13, 145].

Thus, scheduling addresses the allocation of limited resources to tasks over time. Resources may be machines in a workshop, runways at an airport and crews at a construction site, processing units in a computing environment and so on. The tasks may be operations in a production process, take-offs and landings at an airport, stages in a construction project, executions of computer programs and so on.

Scheduling is an optimisation process, and thus refers to one or more **goals** (also called **objectives**, or **performance criteria**). Typical performance criteria in a manufacturing context are:

- the **minimization of the completion time** for the last task;
- the **minimization of the number of tasks completed** after the committed due dates;
- the **throughput** (Q), defined as the number of orders the manufacturing system finishes per time unit;
- the **work-in-process** inventory (WIP), defined as the number of orders in the system which are not finished yet;
- the mean order **flow time** (or **lead time**, $\bar{F}$), defined as the difference between the order finish time and the order start time);
- the mean order **tardiness** $\bar{T}$ (The tardiness is the difference between the order finish time and the due date, if this difference is positive. Otherwise, the tardiness is 0).

The resource allocation process is usually subject to constraints. In practical situations, not all possible resource allocation decisions are feasible. Typical constraints express the limited capacity of a resource, or precedence relations between operations.

Defining performance metrics allows the team to answer a question: How do we tell a good schedule from a bad schedule? Manufacturers should expect some improvement in total system performance, (profitability, throughput, etc.) from implementation of a scheduling process. Additionally, a manufacturer's expectations should be defined and measurable. Typical first-pass requirements for a scheduling process range from "need to know if we can get the work done on time" to "minimize total manufacturing costs." Scheduling team members who try to refine these requirements will run into tough questions, such as, how do we measure the cost of delivering one day late to the customer? Job shop-type operations such as tool and die facilities or prototype shops will often be more due-date driven, and end up with requirements that include "minimize maximum lateness," or "no late orders." Batch facilities more frequently build to stock, and will have other requirements. "Minimize total cost of inventory and setups" and "maximize throughput of bottleneck operations" are some typical examples of those types of requirements. Batch facilities that build components for assembly typically require "minimize inventory costs without allowing a shortage to occur." In any case, "feasibility" as a performance criteria may be acceptable for a first scheduling implementation. Once the scheduling process is in place, the structure will be available for a continuous improvement effort, and methods will be in place to identify additional opportunities to improve profitability. In addition to definition of the performance metrics, the team must also define requirements for the scheduling system itself. Questions that must be answered in this step include:

- What is an acceptable scheduling lead time?
- What type of planning horizon will we use?
- Do we expect long-term capacity planning, as well as daily dispatch lists?
- What annual budget is available to maintain the scheduling process?
- Are we a candidate for an optimization technique, or will we perform "what if analysis" each time we need to build a schedule?
- How much manual data entry is acceptable?

When facing these questions, one should remember that there is no need to plan in any greater detail than his ability to react to a problem. If process cycle times are in terms of hours, then a daily schedule is usually appropriate and schedule generation time should be less than an hour.

We recall that there is a standard 3-fields notation that identifies each scheduling problem; this triple $\alpha|\beta|\gamma$ is used to report the execution environment, e.g., flow shop, job shop, etc., the various constraints on the job, and the objective function to be optimized, respectively [43].

1.3.2 Mathematical Model for Job Shop Scheduling

For a clear understanding of scheduling and manufacturing control, it is often helpful to define a mathematical model of the scheduling problem. Such a model is heavily dependent on the specific manufacturing system focused on. Therefore, this subsection defines a mathematical model for (extended) job shop scheduling that is directly related to the problem under study in this book. The model can be represented as follows.

To define the symbols, suppose the manufacturing system consists of M workstations (machines) m_k, with k ranging from 0 to $M-1$. Let Ω denote the set of workstations. There are also A auxiliary resources denoted as r_a in the system, with a ranging from 0 to $A-1$. If necessary, these auxiliary resources can be distributed among the workstations.

Consider the set of orders O, with N orders i, for i ranging from 0 till $N-1$. Each order i has:

- a release date R_i, denoting the time an order enters the system;
- a due date D_i, denoting the time the order should be ready;
- a weight w_i, denoting the importance of an order.

Each order i has a number of operations (i,j), for j ranging from 0 to $N_i - 1$. Each operation (i,j) can be executed on the following set of alternative workstations:

$$H_{ij} = \{m_k \in \Omega | m_k \text{ can execute operation } (i,j)\},$$

If operation (i,j) can be executed on workstation k (i.e., $m_k \in H_{ij}$), it has a duration d_{ijk}, otherwise, d_{ijk} is undefined. If the duration is the same for all alternative workstations, d_{ijk} can be simplified to d_{ij}.

Given these data, the resource allocation process selects which workstation $m_k \in H_{ij}$ will execute operation (i,j) and at which time b_{ij} it will start. In other words, $m_k \in H_{ij}$ and b_{ij} are the decision variables. c_{ij} is the resulting completion time of operation (i,j):

$$c_{ij} = b_{ij} + d_{ijk},$$

where d_{ijk} is short for d_{ijm_k}. C_i is the order completion time:

$$C_i = c_{i,N_i-1}.$$

The resource allocation problem is subject to capacity constraints and precedence constraints. Capacity constraints establish that a machine can only execute one operation at the same time. To introduce a mathematical formulation of the capacity constraints, Hoitomt [96] and Luh [125] use a formulation with discretised time t and Kronecker deltas:

$$\sum_{ij} \delta_{ijtk} \leq M_{tk}, \quad (t=1,\ldots,T; \ k=0,1,\ldots,M-1),$$

where M_{tk} is the capacity of workstation m_k at time t, and $\delta_{ijtk} = 1$ if operation (i,j) occupies workstation m_k at time t, and otherwise $\delta_{ijtk} = 0$. T is the time horizon.

Precedence relations can be modelled with sequences of operations [145], with precedence graphs [97, 145] (see Figures 1.18-1.19), simple fork-join precedence constraints [125], assembly trees [117], dynamic precedence graphs [168] or non-linear process plans [63].

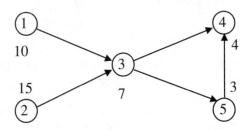

Fig. 1.18. Example of precedence graph

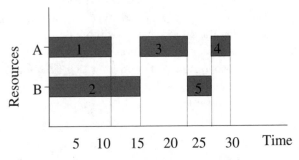

Fig. 1.19. Schedule of the precedence graph in Figure 1.18 on two resources

The experienced reader may notice that this representation of the scheduling problem still does not cover all situations and does not model all possible sources of flexibility in full detail. For instance, and-or graphs can better model choices among processes that imply different sets of operations. For machines with multiple capacity, the exact modelling of set-ups and loading becomes more complex. However, a complete modelling of the scheduling problem becomes too complex to address in a generic way and would be too problem specific to describe in this book.

The model of the scheduling problem described in this book is specific for a certain type of distributed and flexible environment, where there is a huge set of jobs that must be processed by a set of resources. Each job can be processed on each distributed resource, and there could be precedence constraints between jobs.

A feasible schedule is called **active schedule** if it is not possible to construct another schedule by changing the order of processing on the machines and having at least one operation finishing earlier and no operation finishing later.

A feasible schedule is called **semiactive schedule** if no operation can be completed earlier without changing the order of processing on any one of the machines.

A **non-delay schedule** is a schedule where no workstation remains idle if an executable operation is available. In other words, if in a non-delay schedule, a workstation is not occupied on time t, then all operations that can be executed on that workstation are scheduled before time t, or cannot be scheduled on time t because of precedence constraints. Dispatching rules can only produce non-delay schedules. For regular performance measures, the optimal schedule is always an active schedule. Non-delay schedules are easier to calculate, but make no statement on optimality.

1.4 On-line Manufacturing Control

Dispatching is the implementation of a schedule, taking into account the current status of the production system [32].

This definition takes the existence of a scheduler for granted. The dispatcher has to execute the schedule, reacting to disturbances as quickly as possible. Therefore, the dispatcher has no time to optimise his decisions. The task of a dispatcher seems to be straightforward, to the extend that usually only simple heuristics are used for decision taking.

A more general term is **on-line manufacturing control**. As suggested by its name, on-line manufacturing control refers to that part of manufacturing control that takes immediate decisions. Traditionally, it is a synonym for dispatching. However, some new manufacturing paradigms that do not use scheduling, require a more general term that is not based on scheduling. Moreover, in the manufacturing world, dispatching refers to simple decision making, while on-line manufacturing control also refers to more advanced methods.

However, due to the $\mathcal{NP}$-complete nature (see the next subsection for details) of resource allocation, dispatching in manufacturing quickly acquired the meaning of applying simple priority rules. Using an analogy, on-line manufacturing control is the task of a foreman. Similar to dispatching, on-line manufacturing control has to react quickly to disturbances, so it cannot optimise performance with time-intensive computations.

On-line manufacturing control also covers some more practical functions. It includes the management, downloading and starting of NC-programs on due time. While Bauer [32] defines monitoring as a separate module, data acquisition and monitoring are often performed by the on-line (manufacturing) control system.

So, the more data needed, the more complex the system the OCM must manage.

Finally, note that on-line manufacturing control does not deal with process control, such as adjusting the parameters in the machining process or calculating the trajectory of a robot.

1.4.1 Computational Complexity of Scheduling and Resource Allocation

The best-known difficulty of the scheduling and resource allocation problems is their computationally hard nature. As the size of the problem grows, the time required for the computation of the optimal solution grows rapidly beyond reasonable bounds. Formally speaking, most scheduling problems are $\mathcal{NP}$-complete. An $\mathcal{NP}$ problem – *nondeterministically polynomial* – is one that, in the worst case, requires polynomial time in the length of the input to obtain a solution with a non-deterministic algorithm [73].

Non-deterministic algorithms are theoretical, idealised programs that somehow manage to guess the right answers and then show that they are correct. An $\mathcal{NP}$-complete problem is NP and at least as hard as any other $\mathcal{NP}$ problem.

$\mathcal{NP}$-completeness proofs are available for a number of simple scheduling problems and realistic problems tend to be even more complex. The practical consequence of $\mathcal{NP}$-completeness is that the required calculation time needed to find the optimal solution grows at least exponentially with the problem size.

In other words, where finding an optimal solution for realistic scheduling and resource allocation problems could require thousands of centuries, it is useless to consider optimal algorithms. This implies that *near optimal* algorithms are the best possible ones. A near optimal algorithm spends time on improving the schedule quality, but does not continue until the optimum is found.

This scenario naturally motivates the algorithmic approaches chosen in this book.

1.5 Algorithmic Approach to Problem Solution

From a practical point of view, a manager dealing with the problems recalled in the previous sections has a to cope with very inefficient situations, where a better operative procedure could be found at hand by just relying on his own experience. This happens in all the cases where the production system, though well designed for a given operative configuration, has not been progressively updated to adapt to external changes induced by market requirements or new technological solutions. An entirely different scenario is when a manager has to improve a manufacturing system which appears well designed and efficient, and yet, because of competing firms in the market, improvements in the current procedure must be found. In this situation, expertise alone is not enough and optimization techniques are required.

In general, optimization methods for manufacturing problems can fall into two categories: mathematical programming tools and discrete optimization algorithms. The former are always powerful tools for problem modeling and representation. They can be utilized in the prototyping phase and to obtain a solution to small size instances by using one of the commercially available solvers, but they usually fail to find solutions to medium or large size problems.

Optimization algorithms for this kind of combinatorial problem (such as scheduling), can be seen as an ad-hoc procedure and can be broadly classified into exact methods and heuristics. Exact algorithms, though ad-hoc designed, have in common with mathematical programming models, limited capabilities in problem size and, in most cases, the time spent to find a certified optimal solution is usually not acceptable when compared to the improvement in the solution quality [13]. Thus, heuristic approaches should be adopted for real size problems. A thorough classification of the techniques in this area is beyond the scope of this book and the reader may refer to [61] also for surveys on recent results.

The algorithms proposed in this book fall in within the broad class of metaheuristics and their characteristics and experimental performance results are described in detail within each chapter.

In this following, we introduce the reader to the different types of algorithms that are contained in the book by means of a very simple example derived from the Multiprocessor Tasks Scheduling Problem (MTSP)(see for example [37, 43]).

The MTSP is defined as follows: there are m identical processors and a list L of n non preemptable tasks (*i.e.*, a task cannot be interrupted during its processing) of different length and we want to find a feasible schedule S of minimum length.

1.5.1 Greedy Heuristics

Greedy algorithms are based on a very intuitive idea, built a solution incrementally, by starting from an empty one and adding an element at a time so

that the cost of this addition is minimum. In our example this can be done as
reported in Table 1.2.

Table 1.2. Greedy Algorithm for MSTP

Input: A list L of n non preemptable tasks of different lengths,
set of m identical processors;
Output: A schedule S that is assignment of tasks to processors
and a starting time for each task;
1. Let $S = \emptyset$;
2. While $L \neq \emptyset$
    2.1. Take the first task i form L and assign it to the
machine with smaller completion time (break ties arbitrarily);
    2.2 Remove i from L.

Let us assume a scheduling system with 2 identical resources and a list L
(T_1, T_2, T_3, and T_4) of 4 tasks of length (listed in the same order tasks are
stored in the list)

$$2, 4, 3, 5,$$

respectively.

The schedule obtained applying the greedy algorithm for MSTP is given
in Figure 1.20.

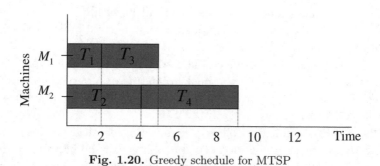

Fig. 1.20. Greedy schedule for MTSP

Clearly, it is not an optimal schedule.

1.5.2 Local Search Heuristics

To override the above problem it appears quite reasonable to improve it by
making same changes in the solution. This concept leads to the definition of a

neighborhood of a solution as a set of feasible solutions that can be obtained from a given (starting) one by making a sequence of changes. In doing so, we measure all the solutions in the neighborhood and choose the best one which is *locally optimal*.

Following our example, let be given a feasible schedule S, let $f(S)$ be its cost (in this specific case its length). The local search algorithm is sketched in Table 1.3.

Table 1.3. Local search algorithm for MTSP

Input: A schedule S;
Output: A schedule S';
1. Let best $= +\infty$;
2. Define $X(S)$ as the set of all feasible schedules obtained from S
  by exchanging two tasks assigned to two different processors;
3. While $X(S) \neq \emptyset$ and $f(S) \leq$ best, do
  3.1 Let best $= f(S)$;
  3.2 Let $S' \in X(S)$ such that $f(S') \leq f(S'')$ for each $S'' \in X(S)$;
  3.3 If $f(S') \leq$ best
    3.3.1 Let $S = S'$;
    3.3.2 Let best $= f(S')$.

When applied to our case, using as input the schedule previously obtained by the greedy algorithm, one get the schedule represented in Figure 1.21.

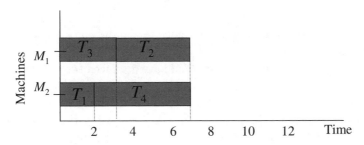

Fig. 1.21. Schedule for MTSP obtained by local search

It is easy to see that this schedule is optimal (by saturation of resource usage on both machines). In general, we might not be always so lucky and the local optimum we get can be far from the global optimum. This, translated into practice, means that a solution obtained by a local search algorithm can be very expensive for the production system. Thus, we further improve the search technique and reach the area of metaheuristic algorithms. The gain in the solution value permitted by a more sophisticated and complex technique can be observed by the experimental results.

1.5.3 Off-line and On-line Algorithms

A further concept is that of off-line algorithm versus on-line algorithms and that of competitive ratio. To simply illustrate the idea, examine again the schedule obtained by the local search algorithm for MTSP (see Figure 1.21).

Assume now that an unexpected task of length 7 is added to the list. Assuming that the schedule has been already delivered to the shop, the only thing we can do is to add this unexpected task to one of the two resources obtaining the schedule in Figure 1.22.

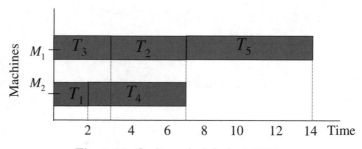

Fig. 1.22. On-line schedule for MTSP

The problem here is that we did not know in advance of the arrival and the duration of a new task (we are in the on-line case). Of course, if we had known this (in an off-line case) we would have produced the schedule in Figure 1.23.

The on-line algorithm is penalized by the fact that a decision taken cannot be changed when a new tasks arrives in the system, while the off-line algorithm can optimize with complete information and obtain a better solution. This is just to introduce the concept of performance ration of an on-line solution over on off-line one. This ratio is computed (virtually) over all instances of the problem and its maximum is called competitive ratio.

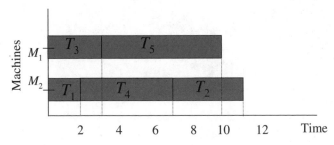

Fig. 1.23. Off-line schedule for MTSP

1.6 Conclusions

In this chapter we have recalled same basic definitions of manufacturing systems and introduced the reader to the areas of the problems studied in this book. We refer the reader to the books in [13, 45] for general manufacturing concepts. Referring to the scheduling models, we suggest [37, 43, 145]. Furthermore, we have explained the reasons that motivate the design of new algorithms to solve these problems. For a general introduction to combinatorial optimization and algorithm complexity the reader can see [73, 142]. Indeed, most of the procedures contained in this book rely on innovative techniques in the design of metaheuristics. In our opinion, simpler algorithms can be used, but one must be ready to accept inferior quality solutions. Hence, the choice depends on the manager's evaluation of both the effort needed to introduce better optimization tools in the production process and on the economic advantage one can obtain from them, depending on the volume of items produced and other factors (time horizons, competition, market demands, *etc.*). To allow the user to judge for himself, each chapter contains extensive computational results on the performance of the various algorithms and a comparative analysis of solutions.

2

On-Line Load Balancing

Without proper scheduling and resource allocation, large queues at each processing operation cause an imbalanced production system: some machines are overloaded while some are starved. With this perspective, the mean cycle time will rise due to local starvation even though the total system inventory stays approximately the same, and it will rise due to the waiting time in the queue for those overloaded processing machines. Hence, the goal is to balance the production facilities to reduce work in progress variability. Thus, production managers are asked to find an allocation strategy so that the workload among the production facilities is distributed as fairly as possible. In other words, managers would like the workload of the different facilities to be kept as balanced as possible. This problem is known as load balancing and is the main topic of this chapter.

2.1 Problem Definition

Formally, the load balancing problem is defined as the problem of the on-line assignment of tasks to n machines; the assignment has to be made immediately upon the arrival of the task, increasing the load on the machine the task has been assigned to for the duration of the task. We consider only nonpreemptive load balancing; $i.e.$, the reassignment of tasks is not allowed. The goal is to minimize the maximum load, or, equivalently, to minimize the sum of the absolute values of the difference between the load of each machine and the average load of the system.

Two main analyzes can be done in this direction: the first one is related to the so called "temporary tasks" problem and the other is the so called "permanent tasks" problem.

The first problem refers to the case in which tasks have a finite duration, $i.e.$, during task arrivals one can observe also task departures from the machines. In the latter problem, tasks are "permanently assigned" to machines, *i.e.*, only task arrivals are observed over time. This can also be interpreted as a problem in which the task duration is very large with respect to the time horizon where tasks arrive over time.

The other type of distinction can be made according to the duration of the task. In fact, we can have either known duration scenarios or unknown duration scenarios.

The on-line load balancing problem naturally arises in many applications involving the allocation of resources. In particular, many cases that are usually cited as applications for bin packing become load balancing problems when one removes the assumption that the items, once "stored", remain in the storage forever. As a simple concrete example, consider the case where each "machine" represents a plant and a task is a work order. The problem is to assign each incoming order to one of the plants which is able to process that work order. Assigning an order to a plant increases the load on this plant, *i.e.*, it increments the percentage of the used capacity. The load is increased for the duration of the request. Formally, each arriving task j has an associated load vector, $p_j = \{p_{1j}, p_{2j}, \ldots, p_{nj}\}$, where p_{ij} defines the increase in the load of machine i if we were to assign task j to it. This increase in load occurs for the duration d_j of the task.

We are interested in the algorithmic aspect of this problem. Since the arriving tasks have to be assigned without knowledge of the future tasks, it is natural to evaluate performance in terms of the competitive ratio. For this problem, the competitive ratio is the supremum, over all possible input sequences, of the maximum (over time and over machines) load achieved by the on-line algorithm to the maximum load achieved by the optimal off-line algorithm. The competitive ratio may depend, in general, on the number of machines n, which is usually fixed, and should not depend on the number of tasks that may be arbitrarily large. Similar to the way scheduling problems are characterized, load balancing problems can be categorized according to the properties of the load vectors.

2.2 Known Results and Existing Approaches

On-line load balancing has been widely investigated in terms of approximation algorithms (e.g., see [14, 16, 17, 18, 19, 20, 21, 115, 144]). The simplest case is where the coordinates of each load vector are equal to some value that depends only on the task. It is easy to observe that Graham algorithm [80], applied to this kind of load-balancing problem, leads to a $(2-(1/n))$-competitive solution. Azar *et al.* [17] proposed studying a less restricted case, motivated by the problem of the on-line assignment of network nodes to gateways (see also [21]). In this case, a task can represent the request of a network node to be assigned to a gateway; machines represent gateways. Since, in general, each node can be served by only one subset of gateways, this leads to a situation where each coordinate of a load vector is either ∞ or equal to a given value that depends only on the task. In this case, which we will refer to as the assignment restriction case, the paper in [17] shows an $\Omega(\sqrt{n})$ lower bound on the competitive ratio of any load balancing algorithm that deals with the unknown duration case, *i.e.*, the case where the duration of a task becomes known only upon its termination. The same authors also present an $O(n^{2/3})$-competitive algorithm. The work in [17] opens several new research ideas. The first is the question of whether there exists an $O(\sqrt{n})$-competitive algorithm for the assignment restriction case when the duration of a task becomes known only upon its termination. Secondly, the $\Omega(\sqrt{n})$ lower bound for the competitive ratio in [17] suggests considering natural variations of the problem for which this lower bound does not apply. One such candidate, considered in this chapter, is the known duration case, where the duration of each task is known upon its arrival.

All the results in this chapter, as well as in the papers mentioned above, concentrate on nonpreemptive load balancing; *i.e.*, the reassignment of tasks is not allowed. Another very different model, is when reassignment of existing tasks is allowed. For the case where the coordinates of the load vector are restricted to 1 or ∞, and a task duration is not known upon its arrival, Phillips and Westbrook [144] proposed an algorithm that achieves an $O(logn)$ competitive ratio with respect to load while making $O(1)$ amortized reassignments per job. The general case was later considered in [16], where an $O(logn)$- competitive algorithm was designed with respect to a load that reroutes each circuit at most $O(logn)$ times. Finally, we note that the load balancing problem is different from the classical scheduling problem of minimizing the makespan of an on-line sequence of tasks with known running times see [81, 157] for a survey. Intuitively, in the load balancing context, the notion of makespan corresponds to maximum load, and there is a new, orthogonal notion of time. See [14] for further discussion on the differences.

2.2.1 The Greedy Approach

When speaking about on-line algorithms, the first that come to mind is the greedy approach. In fact, it is straightforward to allocate an incoming task to the lightest machine. In more detail, one can do as follows:

1. Let $f(s^{(t)})$ be a function defining the maximum load over all the machines at time t, *i.e.*, it returns the weight of the heaviest machine in solution $s^{(t)}$.

2. When a task arrives, let $N(s^{(t)})$ be the neighborhood of current solution $s^{(t)}$;

3. Choose the best solution in $N(s^{(t)})$, *i.e.*, the one that produces the smallest increase of $f(s^{(t-1)})$.

Note that in Line 3 there could be many ties when there is more than one machine for which either the incoming task does not produce any increase in $f(s^{(t-1)})$, or the same smallest positive increment in $f(s^{(t-1)})$ is achieved. In this case one can choose at random the machine in the restricted neighborhood formed exclusively by these solutions.

The greedy algorithm is also known as the Slow-Fit algorithm; it is essentially identical to the algorithm of Aspnes, Azar, Fiat, Plotkin and Waarts for assigning permanent tasks [14]. Roughly speaking, the idea (which originated in the paper by Shmoys, Wein, and Williamson [157]) is to assign the task to the least capable machine while maintaining that the load does not exceed the currently set goal. However, the analysis in [14] is inapplicable for the case where tasks have limited duration. It is known that Slow-Fit is 5-competitive if the maximum load is known. The Slow-Fit algorithm is deterministic and runs in $O(n)$ time per task assignment.

Although this is the simplest way to proceed, both in terms of implementation and in terms of computing times, the lack of knowledge about future incoming tasks can produce the effect of obtaining good solutions in the first algorithm iterations, but may return very low quality solutions after these initial stages.

Thus, one can improve the greedy algorithm by using semi-greedy strategies. At each construction iteration, the choice of where the incoming task has to be allocated is determined by ordering all the candidate elements (*i.e.*, the machines) in a candidate list C with respect to a greedy function $g : C \to R$. This function measures the (myopic) benefit of selecting such an element. In our case g corresponds to f and C contains all the solutions in the neighborhood. Thus, contrary to what the greedy algorithm does, *i.e.*, select the first element in list C which locally minimizes the objective function, the semi-greedy algorithm randomly selects an element from the sublist of C formed by the first r elements, where r is a parameter ranging from 1 to C. This sublist, formed by the first r elements of C, is denoted as the Restricted Candidate List (RCL).

It is easy to verify that if $r = 1$ then the semi-greedy algorithm becomes a greedy algorithm, and if $r = |C|$ then the semi-greedy algorithm behaves like a random algorithm, *i.e.*, a machine is chosen at random and the incoming task is allocated to this machine regardless of how large the increase in f becomes.

For the construction of the RCL, considering the problem of minimizing the maximum load over all the machines at time t, we denote as $\Delta(s_i^{(t)})$ the incremental load associated with allocating the incoming task to machine i, i.e., with the solution $s_i^{(t)} \in N(s^{(t-1)})$ under construction. Let Δ_{min} and Δ_{max} be, respectively, the smallest and the largest increment.

The restricted candidate list RCL is made up of elements $s_i^{(t)}$ with the best (i.e., the smallest) incremental costs $\Delta(s_i^{(t)})$. This list can be limited either by the number of elements (cardinality-based) or by their quality (value-based).

In the first case, it is made up of the r elements with the best incremental costs, where r is the aforementioned parameter. In the latter case, we can construct the RCL considering the solutions in the neighborhood whose quality is superior to a threshold value, i.e., $[\Delta_{min}; \Delta_{min} + \alpha(\Delta_{max} - \Delta_{min})]$, where $0 \leq \alpha \leq 1$. If $\alpha = 0$, then we have a greedy algorithm, while if $\alpha = 1$, we obtain a pure random algorithm.

2.2.2 The Robin-Hood Algorithm

In this section we describe a $(2\sqrt{n}+1)$-competitive algorithm that also works with assignment restriction, where the task duration is unknown upon its arrival. Task j must be assigned to one of the machines in a set M_j; assigning this task to a machine i raises the load on machine i by w_j. The input sequence consists of task arrival and task departure events. Since the state of the system changes only as a result of one of these events, the event numbers can serve as time units; i.e. we can view time as being discrete. We say that time t corresponds to the t-th event. Initially, the time is 0, and time 1 is the time at which the first task arrives. Whenever we speak about the "state of the system at time t", we mean the state of the system after the t-th event is handled. In other words, the response to the t-th event takes the system from the state at t-1 to the state at t. Let OPT denote the load achievable by an optimum off-line algorithm. Let $l_i(t)$ denote the load on machine i at time t, i.e., after the t-th event. At any time t, we maintain an estimate $L(t)$ for OPT satisfying $L(t) \leq OPT$. A machine i is said to be rich at some point in time t if $l_i(t) \geq \sqrt{n}L(t)$, and is said to be poor otherwise. A machine may alternate between being rich and poor over time. If i is rich at t, its windfall time at t is the last moment in time at which it became rich. More precisely, i has windfall time t_0 at t if i is poor at time $t_0 - 1$, and is rich for all times t' where $t_0 \leq t' \leq t$.

The Robin-Hood Algorithm.

The Robin-Hood (RH) algorithm is simple, deterministic, and runs in $O(n)$ time per task assignment. Interestingly, its decision regarding where to assign a new task depends not only on the current load on each machine, but also on the history of previous assignments [18, 20].

Assign the first task to an arbitrary machine, and set $L(1)$ to the weight of the first task. When a new task j arrives at time t, set:

$$L(t) = \max\{L(t-1), w_j, \mu(t)\}$$

The latter quantity, $\mu(t)$, is the aggregate weight of the tasks currently active in the system divided by the number of machines, i.e., $\mu(t) = \frac{1}{n}(w_j + \sum_i l_i(t-1))$. Note that the recomputation of $L(t)$ may cause some rich machines to be reclassified as poor machines.

The generic steps of the RH algorithm are the following:

1. If possible, assign j to some poor machine i.
2. Otherwise, j is assigned to the rich machine i with i being the most recent windfall time.

Lemma 1. *At all times t, the algorithm guarantees that $L(t) \leq OPT$.*

Proof. The proof is immediate since all three quantities used to compute $L(t)$ are less than or equal to OPT.

The following lemma is immediate since $nL(t)$ is an upper bound on the aggregate load of the currently active tasks.

Lemma 2. *At most $\sqrt{n}$ machines can be rich at any point in time.*

Theorem 1. *The competitive ratio of the RH algorithm is at most $2\sqrt{n}+1$.*

Proof. We will show that the algorithm guarantees that at any point in time t, $l_i(t) \leq \sqrt{n}L(t) + OPT) + OPT$ for any machine i. The claim is immediate if i is poor at t. Otherwise, let S be the set of still active tasks at time t that was assigned to i since its windfall time t. Let j be some task in S. Since i is rich throughout the time interval $[t_0, t]$, all the machines M_j that could have been used by the off-line algorithm for j must be rich when j arrives. Moreover, each machine in $M(j) - \{i\}$ must have already been rich before time t, since otherwise, the RH algorithm would have assigned j to it. Let k be the number of machines to which any of the tasks in S could have been assigned; i.e., $k = |\cup_{j \in S} M_j|$. Lemma 2 implies that $k \leq \sqrt{n}$.

Let q be the task assigned to i at time t that caused i to become rich. Since $w_q \leq OPT$, $\sum_{j \in S} w_j \leq kOPT$ and $k \leq \sqrt{n}$ we conclude that

$$l_i(t) \leq \sqrt{n}L(t) + w_q + \sum_{j \in S} w_j \leq \sqrt{n}L(t) + OPT + \sqrt{n}OPT.$$

2.2.3 Tasks with Known Duration: the Assign1 Algorithm

This section considers the case where the duration of a task is known upon its arrival; *i.e.*, d_j is revealed to the on-line algorithm when task j arrives. We describe an algorithm [18, 20] whose competitive ratio is $O(\log nT)$, where T is the ratio of the maximum to minimum duration if the minimum possible task duration is known in advance. The algorithm is based on the on-line algorithm of [14], which solves the following "route allocation" problem: We are given a directed graph $G = (V, E)$ with $|V| = N$. Request i is specified as $(s_i, t_i, \{p_{i,e} | e \in E\})$, where $s_i, t_i \in V$ and for all $e \in E$, $\{p_{i,e}\} \geq 0$. Upon arrival of request i, the route allocation algorithm has to assign i to a path (route) P from s_i to t_i in G; the route assignments are permanent. Let $P = P_1, P_2, \ldots, P_k$ be the routes assigned to requests 1 through k by the on-line algorithm, and let $P^* = P_1^*, P_2^*, \ldots, P_k^*$ be the routes assigned by the off-line algorithm. Given a set of routes P, the load on edge e after the first j requests are satisfied is defined as:

$$l_e(j) = \sum_{i \leq j : e \in P_i} p_{i,e}.$$

Denote the maximum load as $\lambda(j) = \max_{e \in E} l_e(j)$. Similarly, define $l_e^*(j)$ and $\lambda^*(j)$ to be the corresponding quantities for the routes produced by the off-line algorithm. For simplicity, we will abbreviate $\lambda(k)$ as λ and $\lambda^*(k)$ as λ^*. The goal of the online algorithm is to produce a set of routes P that minimizes λ/λ^*. The online route allocation algorithm of [14] is $O(log N)$-competitive, where N is the number of vertices in the given graph. Roughly speaking, we will reduce our problem of the on-line load balancing of temporary tasks with known duration to several concurrent instances of the online route allocation for permanent routes problem, where $N = nT$. Then we will apply the algorithm of [14] to achieve an $O(log N) = O(log nT)$ competitive ratio . We assume that the minimum task duration is known in advance. Let $a(j)$ denote the arrival time of task j, and assume $a(1) = 0$. Let $T' = (\max_j a(j) + d(j))$ be the total duration of all the tasks. First, we make several simplifying assumptions and then show how to eliminate them:

- T' is known in advance to the on-line algorithm.
- T is known in advance to the on-line algorithm.
- Arrival times and task durations are integers; *i.e.*, time is discrete.

We now describe an $O(log nT')$-competitive algorithm called Assign1.

The Assign1 algorithm.

The idea is to translate each task into a request for allocating a route. We construct a directed graph G consisting of T' layers, each of which consists of

$n + 2$ vertices, numbered $1, \ldots, n + 2$. We denote vertex i in layer k as $v(i, k)$. For each layer $1 \leq k \leq T'$, and for $1 \leq i \leq n$, we refer to vertices $v(i, k)$ as common vertices. Similarly, for each layer k, $v(n + 1, k)$ is referred to as the source vertex and $v(n + 2, k)$ is referred to as the sink vertex. For each layer, there is an arc from the source $v(n + 1, k)$ to each of the n common vertices $v(i, k + 1)$. In addition, there is an arc from each common vertex $v(i, k)$ to the sink $v(n+2, k)$ in each layer. Finally, there is an arc from each common vertex $v(i, k)$ to the corresponding common vertex $v(i, k + 1)$ in the next layer. The arc from $v(i, k)$ to $v(i, k + 1)$ will represent machine i during the time interval $[k, k + 1)$ and the load on the arc will correspond to the load on machine i during this interval.

We convert each new task j arriving at $a(j)$ into a request for allocating a route in G from the source of layer $a(j)$ to the sink of layer $a(j) + d(j)$. The loads p_j are defined as follows: for arcs $v(i, k)$ to $v(i, k + 1)$ for $a(j) \leq k \leq a(j) + d(j) - 1$, we set $p_{i,e} = p_i(j)$; we set $p_{j,e}$ to 0 for the arcs out of the source of layer $a(j)$ and for the arcs into the sink of layer $a(j) + d(j)$, and to ∞ for all other arcs. Clearly, the only possible way to route task j is through the arcs $v(i, k)$ to $v(i, k + 1)$ for $a(j) \leq k \leq a(j) + d(j) - 1$ some i. This route raises the load on the participating arcs by precisely $p_i(j)$, which corresponds to assigning the task to machine i. Thus, minimizing the maximum load on an arc in the on-line route allocation on the directed graph G corresponds to minimizing the maximum machine load in the load balancing problem.

We now show how to construct an $O(\log nT)$-competitive algorithm. Partition the tasks into groups according to their arrival times. Group m contains all tasks that arrive in the time interval $[(m - 1)T, mT]$. Clearly, each task in group m must depart by time $(m + 1)T$, and the overall duration of tasks in any group is at most $2T$. For each group, invoke a separate copy of Assign1 with $T' = 2T$. That is, assign tasks belonging to a certain group regardless of the assignments of tasks in other groups. Using the route allocation algorithm of [14], we get that for each group, the ratio between maximal on-line load to the maximal off-line load is at most $O(\log nT)$. Moreover, at each instance of time, active tasks can belong to at most 2 groups. Thus, the maximal on-line load at any given time is at most twice the maximal on-line load of a single group. The off-line load is at least the largest load of the off-line algorithms over all groups, and hence the resulting algorithm is $O(\log nT)$-competitive. We now show how to remove the remaining simplifying assumptions. To remove the restriction that T is known in advance, let $T = d(1)$ when the first task arrives. If we are currently considering the m-th group of tasks, and task j arrives with $d(j) > T$, we set $T = d(j)$ and $T' = 2d(j)$. Observe that this does not violate the invariant that at any point in time the active tasks belong to at most two groups. We use the fact that the route allocation algorithm of [14] is scalable, in the sense that the current assignment of tasks in group m is consistent with the assignment if Assign1 had used the new value of T' instead of the old one. Thus, the argument of $O(lognT)$-competitiveness holds as before. Finally, to eliminate the assumption that events coincide with the clock, that is, for all j, $a(j)$'s and $d(j)$'s are integral multiples of time units, Assign1 approximates $a(j)$ by $\lfloor a(j) \rfloor$ and $d(j)$ by $\lceil a(j) + d(j) \rceil - \lfloor a(j) \rfloor$. Since for all j, $d(j) \geq 1$, this approximation increases the maximal off-line load by at most a factor of 2. Thus, the following claim holds:

Theorem 2. *The above on-line load balancing algorithm with known task duration is $O(lognT)$-competitive.*

2.3 A Metaheuristic Approach

As can be inferred from the previous section, due to the on-line nature of the load balancing problem (as happens for the greater part of on-line problems), the approaches used to optimize the system are one-shot algorithms, *i.e.*, heuristics that incrementally build a solution over time; we recall that no reallocation is allowed. Nonetheless, we can reinterpret the RH algorithm as a sort of Stochastic hill-climbing method and, in particular, as a special class of Threshold Acceptance (TA) algorithm, which belongs to the class of metaheuristics.

The novelty of what we propose in the following stems also from the fact that the algorithm is still an on-line constructive algorithm, and thus guarantees reduced computing times, but acts as a more sophisticated approach, where a neighborhood search has to be made as for a local search method.

The striking difference between RH-like algorithms and metaheuristics lies in the fact that the former do not use an objective function to optimize explicitly, but use thresholds, e.g., see $L(t)$, to distinguish between "more profitable" solutions/choices and "less profitable" solutions/choices. This is why the RH algorithm can be associated with the TA algorithm.

The general framework in which a metaheuristic works is depicted in Table 2.1.

where $s^{(0)}$ is the initial solution and $s^{(t)}$ is the current solution at timestep t.

Stochastic hillclimbing methods escape from local minima by probabilistically accepting disimprovements, or "uphill moves". The first such method, Simulated Annealing (SA), was proposed independently by Kirkpatrick *et al.* [25] and Cerny [6] and is motivated by analogies between the solution space of an optimization instance and the microstates of a statistical thermodynamical ensemble. Table 2.2 summarizes the functionalities of the SA algorithm,

Table 2.1. Metaheuristic template

---

1. Let $s^{(0)}$ be an initial solution; $t = 0$.

2. Repeat until a stopping criterion is satisfied:

    2.1. Find a local optimum s_l with local search starting from $s^{(t)}$.

    2.2. Decide whether $s^{(t+1)} = s^{(t)}$ or $s^{(t+1)} = s^{(t-1)}$.

    2.3. $t = t + 1$.

---

which uses the following criteria for Line 2.2 of Table 2.1. If s' is a candidate solution and the function f has to be minimized, and $f(s') < f(s_i)$, then $s_{i+1} = s'$, i.e., the new solution is adopted. If $f(s') \geq f(s_i)$ the "hill-climbing" disimprovement to $s_{i+1} = s'$ still has a nonzero probability of being adopted, determined both by the magnitude of the disimprovement and the current value of a temperature parameter T_i. This probability is given by the "Boltzmann acceptance" criterion described in Line 3.3 of Table 2.2.

Table 2.2. SA template; max_it is a limit on number of iterations

---

    1. Choose (random) initial solution s_0.

    2. Choose initial temperature T_0.

    3. For $i = 0$ to $max\_it - 1$

        3.1. Choose random neighbor solution $s' \in N(s_i)$.

        3.2. If $f(s') < f(s_i)$ then $s_{i+1} = s'$

        3.3. else $s_{i+1} = s'$ with $Pr = exp((f(s_i) - f(s'))/T_i)$.

        3.4. $T_{i+1} = next(T_i)$.

    4. Return s_i, $0 \leq i \leq max\_it$, such that $f(s_i)$ is minimum.

In contrast to SA, TA relies on a threshold T_i, which defines the maximum disimprovement that is acceptable at the current iteration i. All disimprovements greater than T_i are rejected, while all improvements less than T_i are accepted. Thus, in contrast to the Boltzmann acceptance rule of annealing, TA offers a deterministic criterion as described in Line 2.2 of Table 2.1.

At timestep i, the SA temperature T_i allows hillclimbing by establishing a nonzero probability of accepting a disimprovement, while the TA threshold T_i allows hillclimbing by specifying a permissible amount of disimprovement. Typical SA uses a large initial temperature and a final temperature of zero (note that $T = \infty$ accepts all moves; $T = 0$ accepts only improving moves, i.e., the algorithm behaves like a greedy algorithm). The monotone decrease in T_i is accomplished by $next(T_i)$, which is a heuristic function of the T_i value and

Table 2.3. TA template; max_it is a limit on the number of iterations.

---

1. Choose (random) initial solution s_0.

2. Choose initial threshold T_0.

3. For $i = 0$ to $max\_it - 1$

    3.1. Choose random neighbor solution $s' \in N(s_i)$.

    3.2. If $f(s') < f(s_i) + T_i$ then $s_{i+1} = s'$

    3.3. else $s_{i+1} = s_i$.

    3.4. $T_{i+1} = next(T_i)$.

4. Return s_i, $0 \le i \le max\_it$, such that $f(s_i)$ is minimum.

---

the number of iterations since the last cost function improvement (typically, $next(T_i)$ tries to achieve "thermodynamic equilibrium" at each temperature value). Similarly, implementations of TA begin with a large initial threshold T_0 which decreases monotonically to $T_i = 0$. Note that both SA and TA will in practice return the best solution found so far, $i.e.$, the minimum cost solution among $s_0, s_1, \ldots, s_{max\_it}$; this is reflected in Line 4 of Tables 2.2 and 2.3.

Going back to the analogy between the RH algorithm and Stochastic hill-climbing, we can in more detail associate the former algorithm with a particular class of TA algorithms denoted as Old Bachelor Acceptance (OBA) algorithms.

OBA uses a threshold criterion in Line 2.2 of Table 2.1, but the threshold changes dynamically – up or down – based on the perceived likelihood of it being near a local minimum. Observe that if the current solution s_i has lower cost than most of its neighbors, it will be hard to move to a neighboring solution; in such a situation, standard TA will repeatedly generate a trial solution s' and fail to accept it. OBA uses a principle of "dwindling expectations": after each failure, the criterion for "acceptability" is relaxed by slightly increasing the threshold T_i, see $incr(T_i)$ in Line 3.3 of Table 2.4 (this explains the name "Old Bachelor Acceptance"). After a sufficient number of consecutive failures, the threshold will become large enough for OBA to escape the current local minimum. The opposite of "dwindling expectations" is what we call ambition, whereby after each acceptance of s', the threshold is lowered (see $decr(T_i)$ in Line 3.2 of Table 2.4) so that OBA becomes more aggressive in moving toward a local minimum. The basic OBA is shown in Table 2.4.

Let us now examine what happens if we translate the RH algorithm in the realm of OBA (in Table 2.5 we show the RH algorithm modified in terms of OBA, which we have denoted as OBA-RH).

Table 2.4. OBA template; max_it is a limit on the number of iterations.

---

1. Choose (random) initial solution s_0.

2. Choose initial threshold T_0.

3. For $i = 0$ to $max\_it - 1$

    3.1. Choose random neighbor solution $s' \in N(s_i)$.

    3.2. If $f(s') < f(s_i) + T_i$ then $s_{i+1} = s'$ and $T_{i+1} = T_i - decr(T_i)$

    3.3. else $s_{i+1} = s_i$ and $T_{i+1} = T_i + incr(T_i)$.

4. Return s_i, $0 \le i \le max\_it$, such that $f(s_i)$ is minimum.

The first key-point to be addressed is how to take into account the objective function $f(s_i)$ that is not considered by the RH algorithm, and is an important issue in OBA.

Denote at timestep t the load l of machine i as $l_i^{(t)}$ and let $\mu(t)$ be the average load of the system at time t, i.e., $\mu(t) = (\sum_i l_i^{(t-1)} + w_j)/n$. Moreover, let us define a candidate solution at timestep t as $s^{(t)}$. Thus, we can define

$$f(s^{(t)}) = \sum_i |l_i^{(t)} - \mu(t)|,$$

i.e., the sum of the absolute deviation from the average load of the system of each machine load. Similarly to how we have denoted a solution, let us denote the acceptance threshold at time t as $T^{(t)}$.

Following Line 1 of the algorithm in Table 2.4 we have to choose an initial solution. Solution $s^{(0)}$ in Line 1 of the OBA-RH algorithm is the empty solution, i.e., the one in which all the machines are unloaded.

Without loss of generality, let us assume we are at timestep t, with solution $s^{(t-1)}$ of the previous step $(t-1)$. Following Line 3.1 of Table 2.4, we have to generate the set of neighboring solutions $N(s^{(t-1)})$ from $s^{(t-1)}$. Let us encode a solution in the neighborhood as follows: $s_1^{(t)}$ is the neighboring solution at timestep t that allocates incoming task j of weight w_j on machine 1, $s_2^{(t)}$, similarly, is the solution obtained by charging machine 2, and so on until the n-th solution in which machine n's load is increased by w_j.

Solution $s_i^{(t)}$ can be represented by vector $(l_1^{(t-1)}, \ldots, l_i^{(t-1)} + w_j, \ldots, l_n^{(t-1)})$, where the generic component k represents the load of machine k when j is assigned to such a machine; thus, it is clear that in $s_i^{(t)}$ all machine loads remain the same as in the previous iteration, except for machine i which is increased by w_j. Hence, the neighborhood of the problem is formed by n possible solutions, i.e., the solutions obtainable by the current one adding, respectively, the new incoming task j to each of the n machines. In this case, choosing a

Table 2.5. OBA-RH template; max_it is a limit on the number of iterations

---

1. The initial solution $s^{(0)}$ is the one where all the machines are empty;
set $T^{(0)} = \infty$.

2. For $t = 1$ to $max\_it - 1$

    2.1. Let j be the task arriving at time t.

    2.2. Evaluate the neighboring solutions $s_i^{(t)} \in N(s^{(t-1)})$.

    2.3. If there are poor machines, then choose at random a
neighboring solution s' corresponding to a poor machine;
this will satisfy $f(s') < f(s^{(t-1)}) + T^{(t-1)}$;
set $T^{(t)} \leq \max\{0, f(s') - f(s^{(t-1)})\}$;

    2.4. else choose threshold $T^{(t)} \geq \max\{0, f(s_{\hat{i}}^{(t)}) - f(s^{(t-1)})\}$,
where $\hat{i}$ is a rich machine whose windfall time is the smallest.

    2.5. $s^{(t)} = s^{(t-1)}$.

3. Return $f(s^{(t)})$.

---

solution at random (see Line 3.1 of Table 2.4) means choosing a machine at random and then adding the incoming task to that machine.

For instance, at timestep $t = 1$, the solution $s^{(1)}$ can be chosen among the following n candidate solutions in $N(s^{(0)})$:

$$s_1^{(1)} = (w_j, 0, \ldots, 0),$$

$$s_2^{(1)} = (0, w_j, \ldots, 0),$$

$$\ldots$$

$$\ldots$$

$$s_n^{(1)} = (0, \ldots, 0, w_j).$$

Define a machine $\tilde{i}$ poor if $f(s_{\tilde{i}}^{(t)}) < f(s^{(t-1)}) + T^{(t-1)}$, and rich otherwise. Note that, since $f(s^{(0)})$ is 0 (all the machines are empty) and $\mu(0) = 0$, all the machines at time 1 will be rich if threshold $T^{(0)}$ is greater than $w_j + \frac{w_j}{n}(n-2)$. To let OBA-RH act as RH, we initially set $T^{(0)} = \infty$; in this way, all the machines are initially poor.

Thus, if one, or more than one, poor machine exists, a neighboring solution s' corresponding to a poor machine is chosen at random; according to the definition of a poor machine, this solution would also be accepted by the OBA algorithm. After solution acceptance, the threshold is decreased (see Line 2.3 of Table 2.5) by a quantity $decr(T^{(t)}) = T^{(t-1)} - f(s') + f(s^{(t-1)})$.

If in the neighborhood there is not a solution obeying Line 3.2 of Table 2.4, and, thus, in terms of OBA-RH a poor machine does not exist, we have to keep the previous solution; therefore, in the next iteration, the threshold is raised as done in Line 2.4 of Table 2.5 and the new solution is searched for among the same neighboring solutions (see Line 2.5 in Table 2.5).

In this case, it should appear clear why we have changed the notation, using t rather than i to indicate the timestep of the algorithm: in fact, when a solution is discarded, the algorithm cannot allocate the incoming task, since it has to modify the threshold in such a way that in the next iteration there is more chance of accepting a neighboring solution.

Thus, it could happen that at timestep t the number of tasks processed till t by the algorithm could be lower than t due to a possible rejection of allocation.

Setting the next threshold in the interval

$$f(s_{\hat{i}}^{(t)}) - f(s^{(t-1)}) \leq T^{(t)} \leq f(s_{\hat{\hat{i}}}^{(t)}) - f(s^{(t-1)})$$

where $\hat{\hat{i}}$ is the next rich machine after $\hat{i}$, allows one to select, in the next iteration, among the subset of poor machines that will be created having the same windfall time, and, in the case of a tie, the same minimum objective function increment; thus, if such a choice of the threshold is made, one can directly allocate task j to machine $\hat{i}$ without generating the successive neighborhood. In this case, we are doing exactly what RH does, by means of OBA. Note that to achieve such a condition, we can set the new threshold equal to

$$T^{(t)} = f(s_{\hat{i}}^{(t)}) - f(s^{(t-1)}) + \epsilon$$

where ϵ is a positive sufficiently small quantity, i.e., $\epsilon \leq f(s_{\hat{\hat{i}}}^{(t)}) - f(s_{\hat{i}})$.

This version of OBA-RH, denoted OBA-RH revised, is depicted in Table 2.6.

Remark 1. Note that we have considered only the instants of time related to a task arrival, since a task departure is not a decision stage; it just decreases the machine load by a value w_j if j is the outcoming task.

2.4 Example

In the following, we provide an example of RH and OBA-RH to compare how they work.

Let us assume to have 4 machines and 5 incoming tasks. For the sake of simplicity, let us also assume that the departure dates of the tasks are

Table 2.6. OBA-RH revised template; max_it is a limit on the number of iterations

---

1. The initial solution $s^{(0)}$ is the one where all the machines are empty;
set $T^{(0)} = \infty$.

2. For $t = 1$ to $max\_it - 1$

    2.1. Let j be the task arriving at time t.

    2.2. Evaluate neighboring solutions $s_i^{(t)} \in N(s^{(t-1)})$.

    2.3. If there are poor machines then choose at random a
neighboring solution s' corresponding to a poor machine;
this will satisfy $f(s') < f(s^{(t-1)}) + T^{(t-1)}$;
set $T^{(t)} = \max\{0, f(s') - f(s^{(t-1)})\}$;

    2.4. else choose threshold $T^{(t)} = \max\{0, \min_i\{f(s_i^{(t)}) - f(s^{(t-1)})\}\} + \epsilon$,
where $\hat{i}$ is a rich machine whose windfall time is the smallest.

    2.5. $s^{(t)} = s_{\hat{i}}^{(t)}$.

3. Return $f(s^{(t)})$.

---

all equal to 6 and that tasks arrive one by one at time 1, 2, 3, 4, and 5, respectively. Moreover, assume that the weights are as follows: $w_1 = 2, w_2 = 5, w_3 = 14, w_4 = 14, w_5 = 4$.

Assume task 1 is the first incoming task in the system; since $L(0) = 0$, we have that

$$L(1) = \max\{L(0), w_1, (w_1 + \sum_i l_i(0))/4\} = 2.$$

Thus, all the machines are poor, since their load is initially zero, and $\sqrt{n}L(1) = 4$.

We choose a poor machine at random, say machine 3; therefore, the current load vector is $p_1 = (0, 0, 2, 0)$. When task 2 arrives,

$$L(2) = \max\{L(1), w_2, (w_2 + \sum_i l_i(1))/4\} = 5.$$

Since $\sqrt{n}L(2) = 10$, all the machines are again poor and we can proceed by randomly choosing a machine, *e.g.*, machine 2. Our load vector is now $p_2 = (0, 5, 2, 0)$.

When task 3 enters the system, $L(3) = \max\{5, 14, 21/4\} = 14$, and again all the machines are poor since $\sqrt{n}L(3) = 28$. We then randomly assign task 3 to a machine, say machine 2. The new load vector is $p_3 = (0, 19, 2, 0)$.

In the next iteration, task 4 arrives and $L(4)$ is equal to 14. Again, it is easy to verify that all the machines are poor and let us suppose randomly choosing machine 2, whose load increases to 33.

Now, when task 5 arrives, $L(5) = 14$ and machine 2 is rich. Thus, we have to choose one poor machine, at random, among machines 1, 3 and 4. Let us suppose we choose machine 1 and the load vector p_5 is $(4, 33, 2, 0)$.

Let us now consider the OBA-RH algorithm. Since $T^{(1)} = +\infty$, all the machines are initially poor and thus we can allocate task 1 as we did for the RH algorithm, at random; as in the previous scenario, we choose machine 3. This corresponds to choosing a solution $s' = s_3^{(1)}$ in the neighborhood $N(s^{(0)})$ of $s^{(0)}$. Thus, the current load vector is the same as the load vector p_1 computed before.

Now, following Line 2.3 of Table 2.6, we set $T^{(2)} = \max\{0, f(s') - f(s^{(1)})\} = \max\{0, 3 - 0\} = 3$.

The next task to be considered is task 2. Evaluating the neighborhood of $s^{(1)}$ we obtain the following:

$$f(s_1^{(1)}) = (5 - 7/4) + 7/4 + (2 - 7/4) + 7/4 = 7,$$

$$f(s_2^{(1)}) = 7/4 + (5 - 7/4) + (2 - 7/4) + 7/4 = 7,$$

$$f(s_3^{(1)}) = 7/4 + 7/4 + (7 - 7/4) + 7/4 = 21/2,$$

$$f(s_4^{(1)}) = 7/4 + 7/4 + (2 - 7/4) + (5 - 7/4) = 7.$$

Thus, we have

$$f(s_1^{(1)}) - f(s^{(1)}) = 7 - 3 = 4,$$

$$f(s_2^{(1)}) - f(s^{(1)}) = 7 - 3 = 4,$$

$$f(s_3^{(1)}) - f(s^{(1)}) = 21/2 - 3 = 15/2,$$

$$f(s_4^{(1)}) - f(s^{(1)}) = 7 - 3 = 4.$$

It is easy to verify that all the machines are rich, since $f(s_i^{(1)}) - f(s^{(1)}) \geq T^{(2)}$ for each machine $i = 1, \ldots, 4$.

Thus, we cannot accept any solution, and have to increase the threshold and then repeat the neighborhood search for a new possibly acceptable solution.

The new threshold is

$$T^{(3)} = \max\{0, \min_i\{f(s_i^{(2)}) - f(s^{(2)})\}\} + \epsilon = \max\{0, 7 - 3\} = 4 + \epsilon, \quad (2.1)$$

where, according to our definition of ϵ, we have that $\epsilon \leq f(s_{\hat{i}}^{(t)}) - f(s_{\hat{i}})$, i.e., $\epsilon \leq 15/2 - 4 = 7/2$, and machines $\{1, 2, 4\}$ allow the achievement of $T^{(3)}$. Suppose we set $\epsilon = 7/2$.

Thus, we allocate task 2 to one of the rich machines in the set $\{1, 2, 4\}$ since they all have the same minimum difference among $f(s_i^{(1)})$ and $f(s^{(1)})$.

We choose machine 2, set $s' = s_2^{(1)}$ and $s^{(2)} = s_2^{(1)}$, and the new load vector is $p_2 = (0, 5, 2, 0)$. Note that this choice of $T^{(3)}$ follows the same rationale as the one behind the OBA algorithm since we observe an increment in the threshold with respect to the previous iteration when a solution rejection occurs.

When task 3 arrives, we have the following situation:

$$f(s_1^{(2)}) = (14 - 21/4) + (21/4 - 5) + (21/4 - 2) + 21/4 = 35/2,$$

$$f(s_2^{(2)}) = 21/4 + (19 - 21/4) + (21/4 - 2) + 21/4 = 55/2,$$

$$f(s_3^{(2)}) = 21/4 + (21/4 - 5) + (16 - 21/4) + 21/4 = 43/2,$$

$$f(s_4^{(2)}) = 21/4 + (21/4 - 5) + (21/4 - 2) + (14 - 21/4) = 35/2.$$

Thus, we have

$$f(s_1^{(2)}) - f(s^{(2)}) = 35/2 - 7 = 21/2$$

$$f(s_2^{(2)}) - f(s^{(2)}) = 55/2 - 7 = 41/2,$$

$$f(s_3^{(2)}) - f(s^{(2)}) = 43/2 - 7 = 29/2,$$

$$f(s_4^{(2)}) - f(s^{(2)}) = 35/2 - 7 = 21/2.$$

It is easy to verify that, if in (2.1) we choose $\epsilon \leq 13/2$ then all the machines are rich because $f(s_i^{(2)}) - f(s^{(2)}) \geq T^{(3)}$ for each $i = 1, \ldots, 4$. Since we chose $\epsilon = 7/2$, we cannot accept any of the solutions in the neighborhood, and the next step is to increase the threshold to $T^{(4)} = 21/2 + \epsilon$ with $0 \leq \epsilon \leq 4$ and then accept one solution between $s_1^{(2)}$ and $s_4^{(2)}$. Suppose we select machine 1, i.e., $s' = s_1^{(2)}, s^{(3)} = s'$, and the new load vector is $p_3 = (14, 5, 2, 0)$. Note that, also in this case, this choice of $T^{(4)}$ follows the rationale of the OBA algorithm where we observe an increment in the threshold with respect to the previous iteration when a solution rejection occurs.

When task 4 arrives, we have the following situation:

$$f(s_1^{(3)}) = (28 - 35/4) + (35/4 - 5) + (35/4 - 2) + 35/4 = 77/2,$$

$$f(s_2^{(3)}) = (14 - 35/4) + (19 - 35/4) + (35/4 - 2) + 35/4 = 31,$$

$$f(s_3^{(3)}) = (14 - 35/4) + (35/4 - 5) + (14 - 35/4) + 35/4 = 25,$$

$$f(s_4^{(3)}) = (14 - 35/4) + (35/4 - 5) + (35/4 - 2) + (14 - 35/4) = 21,$$

Thus, we have

$$f(s_1^{(3)}) - f(s^{(3)}) = 77/2 - 35/2 = 21,$$

$$f(s_2^{(3)}) - f(s^{(3)}) = 31 - 35/2 = 13,$$

$$f(s_3^{(3)}) - f(s^{(3)}) = 25 - 35/2 = 15/2,$$

$$f(s_4^{(3)}) - f(s^{(3)}) = 21 - 35/2 = 7/2.$$

It is easy to verify that machines 3 and 4 are poor, whatever the value of ϵ; while machine 2 is poor if $\epsilon \geq 5/2$, and machine 1 is rich regardless of ϵ. Assuming we have set $\epsilon = 0$, we have to choose one poor machine between machines 3 and 4. Suppose we choose machine 4, *i.e.*, $s^{(4)} = s_4^{(3)}$, our load vector is then $p_4 = (14, 5, 2, 14)$ and we set the threshold at

$$T^{(3)} = \max\{0, \{f(s_4^{(3)} - f(s^{(3)}))\}\} = 7/2.$$

Note that, as with the OBA algorithm, the threshold is decreased due to a solution acceptance.

When task 5 arrives, we have the following situation:

$$f(s_1^{(4)}) = (18 - 35/4) + (35/4 - 5) + (35/4 - 2) + (14 - 35/4) = 25,$$

$$f(s_2^{(4)}) = (14 - 35/4) + (9 - 35/4) + (35/4 - 2) + (14 - 35/4) = 175/4,$$

$$f(s_3^{(4)}) = (14 - 35/4) + (35/4 - 5) + (35/4 - 6) + (14 - 35/4) = 17,$$

$$f(s_4^{(4)}) = (14 - 35/4) + (35/4 - 5) + (35/4 - 2) + (18 - 35/4) = 25,$$

Thus, we have

$$f(s_1^{(4)}) - f(s^{(4)}) = 25 - 21 = 4,$$

$$f(s_2^{(4)}) - f(s^{(4)}) = 175/4 - 21 = 91/4,$$

$$f(s_3^{(4)}) - f(s^{(4)}) = 17 - 21 = -4,$$

$$f(s_4^{(4)}) - f(s^{(4)}) = 25 - 21 = 4.$$

It is easy to verify that only machine 3 is poor, and thus, we allocate task 5 to this machine to obtain a final load vector equal to $p_5 = (14, 5, 7, 14)$.

Note that the vector so obtained is the best possible solution obtainable by an off-line algorithm; the overall unbalance of the system at the end of the allocation is equal to

$$(14 - 10) + (10 - 5) + (10 - 7) + (14 - 10) = 16,$$

being that $\mu(5) = 19$.

We conclude noting that, based on the hypothesis made at the beginning of the example, after the arrival of task 5, the system is emptied since all the tasks leave the machines.

2.5 Experimental Results

We have experimented the greedy , the semi-greedy, the RH, and the OBA-RH revised algorithms on random instances with:

- 100, 150, 200, 250, 300, 350, 400, 450 and 500 tasks,
- a number of machines from 5 to 20,
- weights $w_j \in \{1, \ldots, 10\}$ assigned at random to each task j,
- arrival date of the tasks is chosen at random in the time interval $[1, 360]$,
- the duration of a task varies at random from 1 to 10 time units.

All the algorithms were implemented in the C language and run on a PC with 2.8 MHz Intel Pentium processor and 512 MB RAM. In this set of experiments the input to the algorithms is a list of events, *i.e.*, an ordered sequence of arrival dates and departure dates of the tasks, that cannot be exploited in advance by the algorithms due to the on-line nature of the problem. Hence, starting from the first event (that must be an arrival time of a certain task since we assume the system empty), the algorithms in case of an incoming task decide the machine onto which allocate such a task, and simply update the load of a machine when the event to be processed is the departure of a task.

The objective function used to obtain the values given in Tables 2.7-2.10 is $\sum_i |\mu(t) - l_i(t)|$. These are the values produced by the algorithm at the end of its run. Note that the results are given as the superior integer part of the objective function. Moreover, note that the results obtained for the semi-greedy algorithm are achieved by fixing $r = \lceil 0.2 \cdot \text{number of tasks} \rceil$.

Table 2.7. Comparison among different on-line load balancing algorithms. The number of machines is 5

	Greedy	Semi-greedy	RH	OBA-RH revised
# tasks: 100	112	108	95	89
150	148	138	122	102
200	201	189	175	140
250	244	225	202	182
300	302	295	255	221
350	341	312	277	256
400	389	365	299	268
450	412	378	325	272
500	478	452	332	289

Table 2.8. Comparison among different on-line load balancing algorithms. The number of machines is 10

	Greedy	Semi-greedy	RH	OBA-RH revised
# tasks: 100	60	54	47	43
150	74	69	61	51
200	100	99	87	120
250	125	112	101	91
300	154	158	128	110
350	178	160	135	128
400	195	182	150	130
450	202	195	161	135
500	235	225	165	147

Table 2.9. Comparison among different on-line load balancing algorithms. The number of machines is 15

	Greedy	Semi-greedy	RH	OBA-RH revised
# tasks: 100	43	35	33	27
150	44	43	41	31
200	50	48	47	44
250	68	65	61	60
300	87	80	102	74
350	98	90	111	77
400	112	102	115	79
450	150	142	122	85
500	180	178	125	92

Results in the tables highlight the behavior of the four algorithms. The greedy algorithm has the worst performance once all the tasks have been processed. This is not surprising since, as we mentioned in the previous section, the greedy algorithm is able to do better in the first iterations of the algorithm run, while it tends to jeopardize solutions over time. To emphasize this, in Table 2.1 we show the trend of the objective function values over time. It should be noted how the greedy algorithm is able to maintain a good (low) objective function value for the first iterations, while this value grows quickly and is not able to consistently reduce the objective function values.

Table 2.10. Comparison among different on-line load balancing algorithms. The number of machines is 20

	Greedy	Semi-greedy	RH	OBA-RH revised
# tasks: 100	35	31	28	21
150	38	34	29	25
200	50	45	32	27
250	52	44	38	34
300	55	50	47	40
350	62	58	48	44
400	68	62	59	51
450	70	72	59	57
500	85	75	59	68

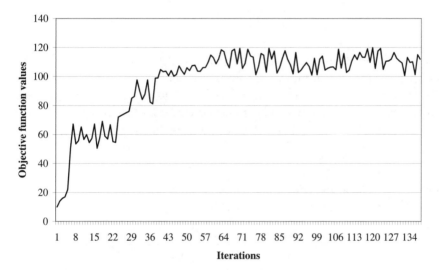

Fig. 2.1. The trend of the objective function of the greedy algorithm over time: the instance with 100 tasks and 5 machines

The semi-greedy algorithm is able to perform better than the greedy algorithm. For the sake of completeness, in Figure 2.2 we show the trend of the objective function values obtained at the end of the algorithm run, varying the values of r, *i.e.*, the cardinality of the restricted candidate list, for the case of 100 tasks and 5 resources. As can be seen, the best values are obtained in correspondence to 0.2, and this justifies our choice of r.

When using the RH and the OBA-RH revised algorithms, we observe a further decrease in the objective function value in the last stage of the

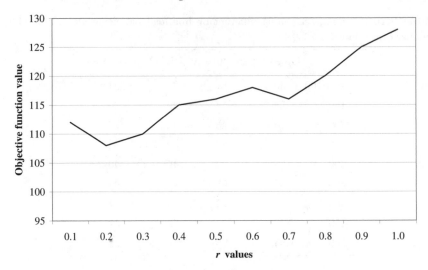

Fig. 2.2. The trend of the objective function of the semi-greedy algorithm

algorithm; in particular, we see that the latter algorithm beats the former, producing results that are up to about 40% better.

To allow a fair comparison, in Tables 2.11-2.14 we showed the average objective function values over the number of iterations. We observe that the general behavior does not change, while the performance gap among the algorithms is enforced.

Table 2.11. Comparison among different on-line load balancing algorithms. The average case with 5 machines

	Greedy	Semi-greedy	RH	OBA-RH revised
# tasks: 100	89.6	86.4	76.0	71.2
150	118.4	110.4	97.6	81.6
200	160.8	151.2	140.0	112.0
250	195.2	180.0	161.6	145.6
300	241.6	236.0	204.0	176.8
350	272.8	249.6	221.6	204.8
400	311.2	292.0	239.2	214.4
450	329.6	302.4	260.0	217.6
500	382.4	361.6	265.6	231.2

Table 2.12. Comparison among different on-line load balancing algorithms. The average case with 10 machines

	Greedy	Semi-greedy	RH	OBA-RH revised
# tasks: 100	49.6	46.4	46.0	41.2
150	72.4	60.4	58.6	41.6
200	80.8	78.2	70.0	65.0
250	92.2	86.0	81.6	75.6
300	120.6	112.0	102.0	89.8
350	135.8	124.6	118.6	104.8
400	165.2	144.0	125.2	114.4
450	178.6	165.4	136.0	117.6
500	200.4	189.6	150.6	131.2

Table 2.13. Comparison among different on-line load balancing algorithms. The average case with 15 machines

	Greedy	Semi-greedy	RH	OBA-RH revised
# tasks: 100	39.6	36.4	36.0	31.2
150	48.4	42.4	37.6	31.6
200	54.8	51.2	46.0	42.0
250	65.2	62.0	61.6	55.6
300	71.6	66.0	70.0	66.8
350	82.8	79.6	71.6	70.8
400	91.2	92.0	89.2	84.4
450	109.6	102.4	100.0	97.6
500	122.4	121.6	115.6	111.2

In Figures 2.3-2.6, we showed the shape of the load of the machines for instances with 200 and 500 tasks, respectively, and 10 machines, for the greedy and the OBA-RH revised algorithms.

2.5.1 A Multi-objective Approach in the Case of Known Task Departure Dates

As for the case of known durations, we can improve the OBA-RH revised algorithm, by considering a new multi-objective function.

In fact, contrary to the previous case, when the duration of a task is known upon its arrival, we can adopt the following objective function:

Table 2.14. Comparison among different on-line load balancing algorithms. The average case with 20 machines

	Greedy	Semi-greedy	RH	OBA-RH revised
# tasks: 100	29.6	26.4	21.0	19.2
150	38.4	35.4	28.6	25.6
200	40.8	38.2	35.0	32.0
250	52.2	46.0	37.6	34.6
300	60.6	52.0	40.0	43.8
350	65.8	56.6	44.6	55.8
400	75.2	65.0	48.2	70.4
450	87.6	77.4	65.0	72.6
500	98.4	85.6	80.6	75.2

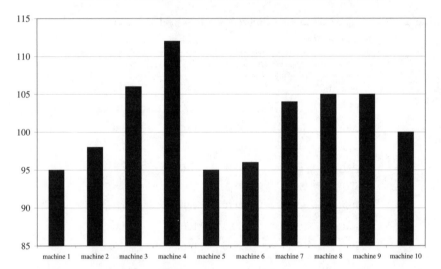

Fig. 2.3. Schedule shape produced by the greedy algorithm on an instance with 200 tasks and 10 machines

$$\min_{i} \alpha \cdot M_i(t) + \beta \cdot \bar{M}_i(t+1, \Delta t)$$

where:

- $M_i(t)$ is the load of machine i once the incoming task is associated with i.
- $\bar{M}_i(t+1, \Delta t)$ is the the average load of machine i in the interval $[t+1, \Delta t]$.
- α and β are two parameters in $[0, 1]$ whose sum is 1.

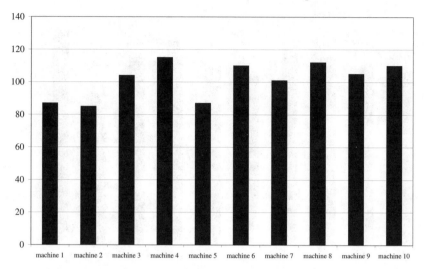

Fig. 2.4. Schedule shape produced by the OBA-RH revised algorithm on an instance with 200 tasks and 10 machines

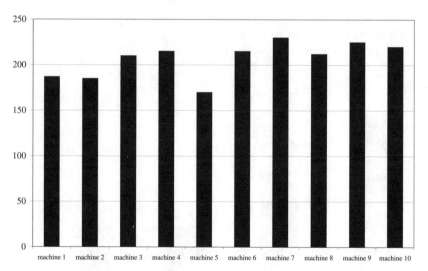

Fig. 2.5. Schedule shape produced by the greedy algorithm on an instance with 500 tasks and 10 machines

Note that in case $\beta = 0$ the objective function reduces to minimize the maximum load.

Example 1. Suppose that the system has three machines and that at time t the incoming task j has weight $w_j = 2$ with duration equal to $d_j = 2$; moreover, assume that assigning task j to machine 1 produces the situation in Figure

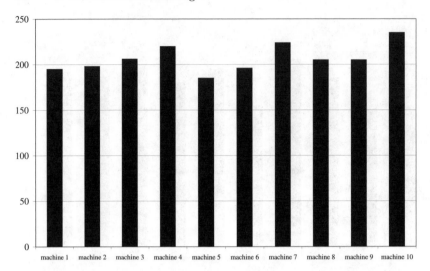

Fig. 2.6. Schedule shape produced by the OBA-RH revised algorithm on an instance with 500 tasks and 10 machines

2.7, and assigning task j to machine 2 produces the situation in Figure 2.8, assigning task j to machine 3 produces the situation in Figure 2.9.

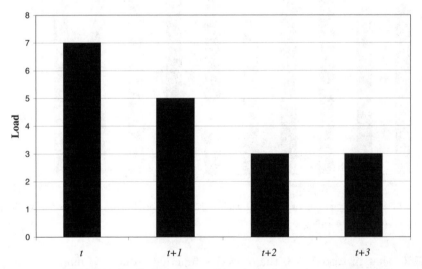

Fig. 2.7. Load shape of machine 1 over time interval $[t, t + 3]$

It is easy to see that if $\Delta t = 2$:

$$\alpha \cdot 7 + \beta \cdot 4$$

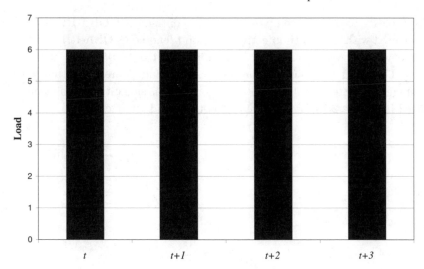

Fig. 2.8. Load shape of machine 2 over time interval $[t, t + 3]$

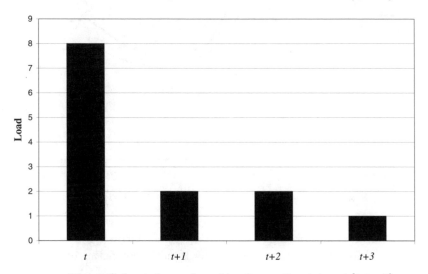

Fig. 2.9. Load shape of machine 3 over time interval $[t, t + 3]$

$$\alpha \cdot 6 + \beta \cdot 6$$

$$\alpha \cdot 8 + \beta \cdot 2.$$

Based on the values of α and β we have a different machine choice. For instance if we have $\beta = 0.8$, we have that the objective function is $\min\{4.6, 6, 3.2\} = 3.2$ and the choice is that of machine 3. On the contrary, if $\beta = 0.2$ then the objective function is $\min\{6.4, 6, 7.6\} = 6$, and the choice is machine 2.

In Tables 2.15-2.18 we compare the results obtained by OBA-RH revised
implemented with the multi-objective function (denoted as OBA-RH$_{rm}$), and
the greedy, semi-greedy, and RH algorithms, with $\alpha = 0.7$, $\beta = 0.3$ and
$\Delta t = 3$. The latter values of the parameters α, β and Δ are those that gave
on average the better results for all the algorithms as suggested by the tuning
reported in Figure 2.10 for the case of 100 tasks and 10 machines.

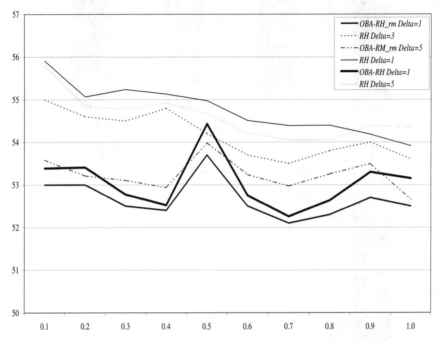

Fig. 2.10. Tuning of the multi-objective parameters

Results on the multi-objective scenario show that OBA-RH$_{rm}$ is able, as
in the single-objective case, to perform better than the competing algorithms.

2.6 Conclusions

In this chapter, we have studied a well known problem in production processes: the balancing of the work load over a set of facilities. This problem
is highly dynamic, that is, in practice, the assignment of an incoming task
to a machine must be done *during* process execution without any knowledge
of future tasks. We proposed and compared four algorithms. Starting from
the simplest greedy heuristic algorithm to a more structured meta-heuristic

Table 2.15. Comparison among different on-line load balancing algorithms with a multi-objective approach. The average case with 5 machines

	Greedy	Semi-greedy	RH	OBA-RH$_{rm}$
# tasks: 100	121.5	120.7	118.2	117.1
150	178.4	176.5	173.4	169.6
200	238.6	236.3	233.6	226.9
250	296.8	293.2	288.8	284.9
300	358.0	356.6	349.0	342.4
350	415.5	409.9	403.2	399.2
400	474.7	470.1	457.4	451.5
450	529.1	522.6	512.4	502.2
500	591.8	586.8	563.7	555.5

Table 2.16. Comparison among different on-line load balancing algorithms with a multi-objective approach. The average case with 10 machines

	Greedy	Semi-greedy	RH	OBA-RH$_{rm}$
# tasks:100	57.7	56.5	55.5	52.1
150	84.4	82.5	81.4	77.1
200	110.1	108.9	107.3	103.3
250	135.9	134.5	133.1	128.8
300	163.0	161.4	159.7	154.5
350	189.2	187.3	185.9	180.2
400	216.4	213.8	211.4	205.7
450	242.4	240.4	238.2	230.9
500	269.0	267.2	263.3	256.6

approach. Improvements in solution quality were very significant in all instances, achieving, for example, almost 40% when comparing the OBA-RH revised algorithm solution to the one produced by the greedy procedure for the instance of 500 tasks and 5 machines.

Table 2.17. Comparison among different on-line load balancing algorithms with a multi-objective approach. The average case with 15 machines

	Greedy	Semi-greedy	RH	OBA-RH$_{rm}$
# tasks: 100	40.3	39.1	38.0	34.9
150	57.6	56.2	54.8	51.6
200	74.8	73.5	72.1	68.8
250	92.2	91.0	90.0	86.1
300	109.4	108.0	107.3	103.3
350	126.9	125.6	124.0	120.2
400	144.2	143.2	142.0	137.6
450	162.2	160.7	159.5	154.9
500	179.8	178.8	177.3	172.2

Table 2.18. Comparison among different on-line load balancing algorithms with a multi-objective approach. The average case with 20 machines

	Greedy	Semi-greedy	RH	OBA-RH$_{rm}$
# tasks: 100	27.2	27.0	26.6	26.0
150	44.4	43.2	41.6	38.8
200	57.1	55.9	54.6	51.6
250	70.4	69.0	67.3	64.2
300	83.5	81.9	80.0	77.2
350	96.4	94.7	92.8	90.3
400	109.6	107.9	107.6	103.5
450	123.1	121.3	119.4	116.1
500	136.4	134.4	133.0	128.8

3

Resource Levelling

In the previous chapter, we examined the problem of balancing the load of a number of production facilities in an on-line scenario where orders (tasks) arriving over time had to be allocated to one of these facilities for processing.

In this chapter we study a more general resource allocation problem in a multiobjective optimization scenario, where the makespan, the peak of resource usage, and the imbalancing of the resource assigned are to be minimized.

The rationale behind this problem, relies on the effect of the parallel execution of activities requiring several resource types. When two tasks are executed simultaneously and overlap for a certain time period (even a small one), the amount of resources of the same type required by both tasks adds up. One may wonder if a schedule where tasks are scheduled to avoid such overlaps exists. This search can be done among schedules with the same makespan, or, according to the decision maker's viewpoint, with a limited increment in the schedule length.

As opposed to load balancing, this problem is typically off-line and each task may require more than one resource at a time. Similarly to the problem studied in the previous chapter, each resource cannot be used simultaneously by more than one task.

3.1 Background and Problem Definition

Many papers dealing with scheduling have faced the problem of how to manage the execution of a set of tasks which must be carried out according to a prefixed resource requirement and a maximum resource availability in the system. A straightforward example is provided by project scheduling (*e.g.*, see [44, 95, 133]). Almost all the research done in this field has focused on minimizing the maximum completion time of all the tasks, respecting the above constraints (*e.g.*, see [34, 35, 44]). This means that, in general, these models do not consider what the resources allocation in the schedule over time, thus revealing their inadequacy in many real life scenarios where a balanced resource distribution is needed to avoid undesired peaks which can be detrimental in terms of costs. The class of problems concerned with scheduling tasks whose objective is to minimize a measure of variation of the resource utilization is known as *resource levelling* or *resource smoothing* scheduling.

This chapter studies the resource levelling problem both in a unitary duration task scenario and arbitrary task duration with incompatibilities, also taking into account the problem of minimizing the maximum completion time. Formally, we can define it is as follows. Given are a set $\mathcal{R} = \{R_1, \ldots, R_K\}$ of K resources types and a set $\mathcal{T} = \{T_1, \ldots, T_n\}$ of n tasks to be carried out using these resources. In particular, the execution of each task $T_i \in \mathcal{T}$ requires the simultaneous availability of a prefixed constant amount of $r_{ik} \geq 0$ units of each resource type R_k, with $k = 1, \ldots, K$. Moreover, incompatibility relationships between pairs of tasks are defined in such a way that when an incompatibility occurs, the tasks involved cannot be scheduled simultaneously. We refer to a schedule S of the tasks in $\mathcal{T}$ as a vector $(t_1, \ldots, t_n)$ of positive integers, where t_i is the starting time of task T_i and $t_i + 1$ is its finishing time. In the following, we will consider only feasible schedules, *i.e.*, those schedules in which tasks respect incompatibilities, and we will refer to the length of S, or the makespan of S, as the maximum completion time of the tasks as scheduled in S.

This setting can be interpreted as a special case of disjunctive scheduling (*e.g.*, see [26, 143]) and of scheduling with forbidden sets as described *e.g.* in [31, 153]. Forbidden sets are used to describe compatibility constraints in scheduling tasks on different types of machines. For instance, if n tasks require the same type of machine, and exactly m machines exist, then at most m of these n tasks can be scheduled simultaneously at any time. This situation can be modelled by forbidden sets, that is subsets of tasks that cannot be scheduled simultaneously. Similarly, our problem can also be modelled as a scheduling problem with forbidden sets, where only unitary duration tasks and forbidden sets with cardinality two are considered.

In order to give some idea of the existing literature on resource levelling, we will now discuss some recent papers on the topic.

In [138], the authors study the problem of levelling resources in project scheduling, where activities are subject to minimum and maximum time lags and explicit resource constraints, *i.e.*, there is a maximum resource availability for each resource type and time period. The time horizon is fixed. The authors propose priority rule heuristics for this problem with a variety of different objective functions. Experiments are presented on instances of up to 500 activities and a comparison with the heuristic presented in [42] is shown.

In [78], the authors consider the problem of optimally allocating resources to competing activities in two different scenarios. In the first one, the resource amounts are given, while in the second one, they are to be determined subject to certain linear constraints. The objective is to find the activity levels such that their weighted deviation from a prespecified target is as small as possible. The authors give a solution algorithm for the first problem whose complexity is of the order of the number of resources times the number of activities, studying also the structure of the optimal value function. Moreover, they provide a pseudo polynomial algorithm for the second problem.

In [139] a resource constrained project scheduling problem with nonregular objective functions is considered. As in [138], minimum and maximum time lags between activities are given. The authors present heuristic and exact algorithms for resource levelling and net present value problems. Experimental results are provided. Other exact algorithms for this problem can be found e.g. in [25, 62, 172].

Algorithms for resource levelling have also been developed in less recent papers by Moder and Phillips in [134], Moder et al. in [135], Ciobanu in [55], and Wiest and Levy in [174], but none of these works give any experimental analysis of the methods proposed, which are mainly based on shifting or priority-rule techniques.

As can be observed from the literature, in the existing models for resource levelling the time horizon is given and one has to manage tasks to achieve objectives related to balancing resource usage in the schedule over time. However, there are cases where one is asked to achieve both well-balanced resource allocation and a minimum makespan, which renders the problem more difficult. In general, we cannot say in advance whether these two objectives are in contrast, meaning that based on the instance, we may have a schedule where the minimum makespan also gives a well balanced schedule, or, on the contrary, a very poor one. What is true, instead, is that if we introduce a further parameter, i.e., the peak of resource usage in the schedule, the latter and the makespan typically exhibit contrary trends (see the following example).

The main contribution of this chapter is to present a novel local search algorithm to deal with both the resource levelling problem and the problem of the minimization of the peak of resource usage and the makespan . An example of an application where considering the three mentioned problems simultaneously may be worthwhile is staff scheduling, where a number of employees have to be assigned to work shifts to carry out a set of activities. In this case, one is mainly concerned with balancing the work schedule of all the employees as well as minimizing the peaks in the schedule, but there could be another issue as well. Indeed, even if the horizon in which the activities have to be scheduled is often given, one may be interested in minimizing the makespan to reserve extra time to manage the risk that one or more activities will exceed their predicted duration.

As will be described in the next sections, the three problems will be managed in a hierarchical way, giving precedence to the minimization of the resource unbalance, and subsequently to the trade-off between the minimization of the peak of resource usage and the makespan.

The remainder of the chapter is organized as follows. In Section 3.2 we discuss the three problems considered; in Section 3.4 we describe the proposed methodology and finally, in Section 3.5 we present our computational experience including a comparison with lower bounds and known algorithms. Algorithms and experimentation in these sections are presented, for the case of unitary (equal) task durations. In Section 3.6 we present the extension of the main algorithm proposed to the case with arbitrary task durations, and, finally, in Sections 3.7 and 3.8, we discuss two case studies.

3.2 Resource Levelling and the Minimization of the Peak and the Makespan

In the next three paragraphs we formally introduce the different objective functions mentioned above, *i.e.*, the makespan, resource balancing and the peak of resource usage, and associate an optimization problem to each one of them. The aim is to tackle simultaneously these problems by means of an optimization algorithm (described in Section 3.4) by imposing a sort of hierarchy on them; in fact, as we said in the previous section, our primary target is to achieve a minimum schedule unbalance, and then to minimize the peak of resource usage and the makespan, paying attention to the proper management of the trade-off between the two latter problems.

The Makespan

The basic problem of minimizing the makespan is to assign tasks to the minimum number of time slots (instants of time) in such a way that there are no conflicts, *e.g.*, pairs of incompatible tasks do not share the same time slots. Let x_{it} be a binary variable indicating if task i is scheduled at t, and let T be an upper bound on the number of available time slots in which all the tasks must be executed. Moreover, let $\mathcal{A}$ be the set of pairs of incompatible tasks. The mathematical formulation associated with the minimization of the makespan is as follows:

$$\min \ \tau \qquad\qquad (3.1)$$

$$\text{s.t.} \ \tau \geq t x_{it}, \quad i = 1, \dots, n, \quad t = 1, \dots, T \qquad (3.2)$$

$$\sum_{t=1}^{T} x_{it} = 1, \quad i = 1, \ldots, n \tag{3.3}$$

$$x_{it} + x_{jt} \leq 1, \quad \forall (i,j) \in \mathcal{A}, \ t = 1, \ldots, T \tag{3.4}$$

$$x_{it} \in \{0,1\}, \quad i = 1, \ldots, n, \ t = 1, \ldots, T \tag{3.5}$$

where (3.1) and (3.2) say that the objective is a min-max function, while (3.3) says that each task must be scheduled in $\{0, \ldots, T\}$ and (3.4) means that if i and j are incompatible they cannot be scheduled in the same time slot. In order to restrain the maximum resource consumption per period, one can also consider the constraints $\sum_{i=1}^{n} r_{ik} x_{it} \leq Q$, $k = 1, \ldots, K$, $t = 1, \ldots, T$ which limit tasks scheduled at time t from using more than a certain amount of Q units of each resource type. It is easy to see that with these constraints the problem is $\mathcal{NP}$-hard, even if constraints (3.4) are neglected as the problem can be reduced to a bin-packing problem [73].

Without resource constraints, the complexity of this problem depends on the degrees of freedom among tasks. In fact, in the absence of incompatibilities (*i.e.*, relaxing constraints (3.4) in the formulation given above) the problem becomes easy and one time slot suffices; otherwise, the problem is the same as that of assigning a label to each task such that no two incompatible tasks receive the same label and the total number of different labels used is minimum. In other words, the problem becomes one of graph coloring, which is $\mathcal{NP}$-hard [73].

Generalizing the problem to arbitrarily task duration, let x_i be a variable associated with the starting time of task i; moreover, let y_{ij} be a binary variables equal to 1 if edge $(i,j) \in \mathcal{A}$ is oriented from i to j and 0 otherwise. Denote the duration of task i as p_i and let M be a very big number. The mathematical formulation associated with the minimization of the makespan is as follows:

$$\min \ \tau \tag{3.6}$$

$$\text{s.t.} \ \tau \geq x_i, \quad i = 1, \ldots, n, \tag{3.7}$$

$$x_i + p_i - x_j \leq y_{ji} M, \quad \forall (i,j) \in \mathcal{A} \tag{3.8}$$

$$x_j + p_j - x_i \leq y_{ij} M, \quad \forall (i,j) \in \mathcal{A} \tag{3.9}$$

$$y_{ij} + y_{ji} = 1, \quad \forall (i,j) \in \mathcal{A} \tag{3.10}$$

$$x_i \in \{0,1\}, \quad i = 1, \ldots, n, \ t = 1, \ldots, T \tag{3.11}$$

$$y_{ij} \in \{0,1\}, \quad \forall (i,j) \in \mathcal{A} \tag{3.12}$$

The objective function is the same as the previous model; constraint (3.8) is such that if edge $(i,j) \in \mathcal{A}$ is oriented from i to j then the right hand side assumes value zero, and we have that

$$x_i + p_i \leq x_j$$

which is a typical longest path constraint; otherwise, *i.e.*, j is oriented towards i, we have that constraint (3.8) is always verified to due the presence of the big M.

Referring to constraints (3.9) we have the opposite situation of constraints (3.8); in fact, when, for an edge (i, j) constraint (3.8) behaves like a longest path constraint, then constraint (3.9) on the same arc is always verified, and, on the contrary, when constraint (3.8) is always verified, then constraint (3.9) is a longest path constraint. Note that this is true because of constraint (3.11) which says that, for edge (i, j), either $y_{ij} = 1$ or $y_{ji} = 1$.

Resource Balancing

Resource balancing can be achieved by minimizing some known objective functions. A first class of objective functions measures the sum of the deviations in the consumption of each resource type from a desired threshold. Letting $\mathbf{tc}^{(t)}$ be the vector whose generic component $tc_k^{(t)}$ represents the resource consumption of type k at period t, an example of these functions is given by the L_1 (or L_2) norm of the difference between $\mathbf{tc}^{(t)}$ and the corresponding desired resource utilization vector. A second type of objective function refers to the so called *resource utilization per period*, defined as the maximum consumption of each resource type in each period. There is also another class of objective functions to be minimized which calculates the sum of the differences of the resource consumptions of consecutive periods. With these functions, one tries to smooth out the amount of resources used from period to period.

In this chapter, we will consider the first two types of objective functions. One is described in the remainder of this subsection, the other is described in the following.

Let S be a schedule of length T. For each time slot $t \leq T$, let the total consumption $tc_k^{(t)}$ of a resource type k be the sum of the resource amounts (of type k) requested by those tasks scheduled at t, *i.e.*,

$$tc_k^{(t)} = \sum_{i=1}^{n} r_{ik} x_{it}.$$

Moreover, let μ_k, with $k = 1, \ldots, K$, be the average resource request (of type k) per period, *i.e.*,

$$\mu_k = \frac{\sum_{i=1}^{n} r_{ik}}{T}$$

and define $dev^{(t)}$, with $t = 1, \ldots, T$, as the average, over the resource types, of the absolute deviations of $tc_k^{(t)}$ from μ_k, that is

$$dev^{(t)} = \frac{\sum_{k=1}^{K} |\mu_k - tc_k^{(t)}|}{K}.$$

Finally, let

$$dev = \sum_{t=1}^{T} dev^{(t)}$$

measure the balancing of the usage of all the resource types over the schedule length; dev is a useful parameter to maintain an efficient resource allocation, as we will show in the next section.

Hence, the mathematical formulation associated with the resource balancing problem within a given time horizon $\{0, \ldots, T\}$ is:

$$\min \ dev = \sum_{t=1}^{T} dev^{(t)} \tag{3.13}$$

$$\text{s.t.} \ \sum_{t=1}^{T} x_{it} = 1, \quad i = 1, \ldots, n \tag{3.14}$$

$$tc_k^{(t)} = \sum_{i=1}^{n} r_{ik} x_{it}, \quad t = 1, \ldots, T, \quad k = 1, \ldots, K \tag{3.15}$$

$$dev^{(t)} = \frac{\sum_{k=1}^{K} |\mu_k - tc_k^{(t)}|}{K}, \quad t = 1, \ldots, T \tag{3.16}$$

$$x_{it} + x_{jt} \leq 1, \quad \forall (i,j) \in \mathcal{A}, \quad t = 1, \ldots, T \tag{3.17}$$

$$x_{it} \in \{0,1\}, \quad i = 1, \ldots, n, \quad t = 1, \ldots, T \tag{3.18}$$

The above problem is $\mathcal{NP}$-hard (see *e.g.* [138]).

The Peak of Resource Usage

Besides dev, another important parameter is the peak of resource usage. Indeed, if on the one hand, one desires a schedule which is perfectly balanced, *i.e.*, with $dev = 0$, on the other hand, one dislikes schedules with a high peak. In this assessment, the schedule length plays a key role since the shorter the schedule the higher the peak, and, similarly, the longer the schedule the lower the peak. To be more specific, suppose we fix a value for the schedule length, say T. Once we are able to find an assignment of tasks to time slots which minimizes dev, we have in effect balanced resource usage over T. Given two schedules with equal dev and different lengths, say T and $T' \neq T$, it is easy to verify that the peak of resource usage is greater in the schedule with $T' < T$. This means that a shorter schedule length can have a better (lower) dev with respect to a longer schedule, while a shorter schedule definitely does not have a better peak than a longer one.

There is also another important aspect which is worth mentioning. In fact, when the number of usable time slots is fixed and a perfect balancing of resources in the schedule is achieved, the system is provided with the minimum quantity of each resource type that one must have to process all the tasks in that number of time slots. Thus, this is also an effective way of determining the amount of resources per period needed to successfully terminate the process.

Let *peak* be the maximum peak in the schedule. The mathematical formulation associated with the minimization of *peak* within a given time horizon $\{0, \ldots, T\}$ is:

$$\min \; peak \tag{3.19}$$

$$\text{s.t.} \; \sum_{t=1}^{T} x_{it} = 1, \quad i = 1, \ldots, n \tag{3.20}$$

$$tc_k^{(t)} = \sum_{i=1}^{n} r_{ik} x_{it}, \quad t = 1, \ldots, T, \quad k = 1, \ldots, K \tag{3.21}$$

$$peak \geq tc_k^{(t)}, \quad k = 1, \ldots, K, \quad t = 1, \ldots, T \tag{3.22}$$

$$x_{it} + x_{jt} \leq 1, \quad \forall (i, j) \in \mathcal{A}, \quad t = 1, \ldots, T \tag{3.23}$$

$$x_{it} \in \{0, 1\}, \quad i = 1, \ldots, n, \quad t = 1, \ldots, T \tag{3.24}$$

When $n \leq T$, the problem is easy since the optimal solution is given by $max_{i \in \mathcal{T}, k=1,\ldots,K} r_{ik}$. Thus, in the following we will consider the case $n > T$.

Note that removing constraints (3.23) does not simplify the problem, even if $K = 1$. In fact, the problem becomes one of scheduling jobs (*i.e.*, tasks) on identical parallel machines (*i.e.*, time slots), with the objective of minimizing the makespan (*i.e.*, *peak*), which is known to be $\mathcal{NP}$-hard [116]. We denote the latter problem as IPMS for short. Note that, in this problem, job i has a processing time equal to $r_{i1} = r_i$.

However, if alongside the relaxation of (3.23) we allow $x_{it} \in [0, 1]$, with $i = 1, \ldots, n$ and $t = 1, \ldots, T$, which in IPMS means that the processing of a job may be interrupted and continued on another machine, then the problem becomes easy. In fact, its optimal solution is given by

$$\left\lceil \sum_{i \in \mathcal{T}} r_{i1} / T \right\rceil = \lceil \mu_1 \rceil.$$

The above expression can be extended to accomplish the case $K > 1$ providing a lower bound on *peak* as

$$LB_{peak}^1 = \max_{k=1,\ldots,K} \lceil \mu_k \rceil.$$

Obviously, relaxing both the integrality constraints on x_{it} and tasks incompatibilities renders LB_{peak} weak when many incompatibilities exist. This phenomenon will be shown later in the analysis of the experimental results.

Further pursuing the connection between our relaxed problem and IPMS, another lower bound can be produced as follows. For ease of presentation assume tasks are numbered according to decreasing processing times, *i.e.*, $r_i \geq r_{i+1}$ for $i = 1, \ldots, n-1$. Consider the $T+1$ largest tasks (we assume n is greater than T) $1, \ldots, T+1$. A lower bound on the minimum makespan for IPMS is given by $r_T + r_{T+1}$, *i.e.*, the sum of the two smallest job times, since at least one machine contains two tasks. In general, a lower bound LB_{peak}^2 is obtained as follows (see *e.g.* [112])

$$ LB_{peak}^2 = \max \left\{ \sum_{i=0}^{j} r_{j \cdot T + 1 - i} \mid j = 1, \ldots, \lfloor (n-1)/T \rfloor \right\}. $$

However, referring to the way we have defined the resource request, given a resource type k, there can exist tasks $i \in T$ such that $r_{ik} = 0$, which in practice risks annihilating the cited lower bound. For this reason, in the experimental analysis, we will use LB_{peak}^1 rather than LB_{peak}^2 for comparison.

Example

Figures 3.1-3.2 depict an instance of our problem with four tasks and three resource types where the pairs of tasks (1,2), (2,3) and (3,4) cannot be processed simultaneously (in the graph, an edge stands for an incompatibility) and tasks require resources as shown in table r_{ik}.

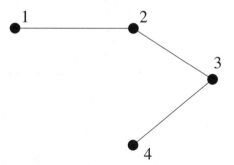

Fig. 3.1. The incompatibility graph

Two different scenarios (schedules) are presented, *i.e.*, S_1, and S_2, in Figure 3.3 and 3.4, respectively. Note that we used the notation $i^{(q)}$ in the schedules, meaning that task i requires q units of the associated resource type (by default

r_{ik}	$k = 1$	2	3
$i = 1$	2	0	1
2	1	2	0
3	1	1	0
4	0	0	1

Fig. 3.2. The resource requirement matrix

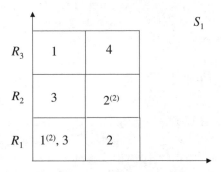

Fig. 3.3. The trade off between the schedule length and the peak of resource usage. The notation $i^{(q)}$ in the schedules means that task i requires q units of the associated resource type. By default $i = i^{(1)}$

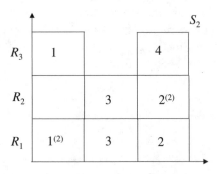

Fig. 3.4. The trade off between the schedule length and the peak of resource usage. The notation $i^{(q)}$ in the schedules means that task i requires q units of the associated resource type. By default $i = i^{(1)}$

$i = i^{(1)}$). Schedule S_1 has length 2, while schedule S_2 occupies 3 time units. Using the parameters defined above, we obtain $dev = 1$ and $peak = 3$ for S_1, $dev = 14/9$ and $peak = 2$ for S_2. These values shed light on the opposing trend of $peak$ with respect to the makespan. Furthermore, it should be noted that by increasing the completion time from 2 to 3, dev goes from 1 to 14/9, but other easy examples can be obtained where an increment of the schedule length corresponds to a reduction in *dev*, meaning that there is no direct relationship between the increase of the makespan and the values assumed by *dev*. This instead suggests examining solutions by means of an algorithmic technique based on the paradigm of local search, *i.e.*, a technique which moves from one solution to another solution, allowing the possibility of worsening one or more of the parameters to be optimized in order to escape from traps or local optima.

In the next section, we describe the algorithmic contribution based on the paradigm of local search.

3.3 The Greedy Approach

The simplest approach is to order tasks according to some kind of rule and then, according to the order found, allocating them to the time slot which locally minimizes the objective function.

Consider the following three parameters associated with the resource requirement:

- Total Resource Requirement (TRR): it is the sum of the resource amounts required by all resource types.

- Maximum Resource Requirement (MRR): it is the maximum resource amount required by all resource types.

- Average Resource Requirement (ARR): it is the average resource amount required by all resource types.

  Consider the following ordering rules:

- Smallest Total Resource Requirement First (STRRF): tasks are ordered according to the smallest total resource requirements.

- Largest Total Resource Requirement First (LTRRF): tasks are ordered according to the largest total resource requirements.

- Alternate Total Resource Requirement(ATRR): tasks are ordered alternating the largest and the smallest total resource requirements.

- Smallest Maximum Resource Requirement First (SMRRF): tasks are ordered according to the smallest maximum resource requirements.

- Largest Maximum Resource Requirement First (LMRRF): tasks are ordered according to the largest maximum resource requirements.

- Alternate Maximum Resource Requirement First (AMRR): tasks are ordered alternating the largest and the smallest maximum resource requirements.

- Smallest Average Resource Requirement (SARRF): tasks are ordered according to the smallest average resource requirements.

- Largest Average Resource Requirement (LARRF): tasks are ordered according to the largest average resource requirements.

- Alternate Average Resource Requirement (AARR): tasks are ordered alternating the largest and the smallest average resource requirements.

Let $\mathcal{O}$ be the set of all the presented ordering rules.

In the case one would like to give more emphasis to the minimization of the makespan, then the simple algorithmic scheme reported in Figure 3.1 should be considered.

Table 3.1. Greedy algorithm (emphasis on makespan)

1. Construct list L, where tasks are ordered according to one of the rules in $\mathcal{O}$.
2. While $L \neq \emptyset$
2.1. Take the first task i from L and assign it to the first feasible time slot.
2.2 Remove i from L.

Note that Line 2.1 in Table 3.1 allocates tasks as soon as possible thus trying to reduce the overall time slots consumption.

In the case one is more interested in the peak minimization, then the greedy algorithm reported in Figure 3.2 should be considered.

Differently from Line 2.1 of Table 3.1, in Table 3.3, Line 2.1 allocates tasks to the time slots which minimize (locally) the peak of resource usage.

For the example in the previous section, we have the following values for the total, maximum and average resource request.

If we consider the rules STRF, LTRF, ATR we obtain the schedule in Figures 3.5-3.7, respectively.

Table 3.2. Greedy algorithm (emphasis on peak)

---

1. Construct list L, where tasks are ordered according to one of the rules in $\mathcal{O}$.
2. While $L \neq \emptyset$
    2.1. Take the first task i from L and assign it to the first feasible time slot which minimizes the peak of resource usage.
    2.2 Remove i from L.

Table 3.3. Total, maximum and average resource requests in the example of the previous section

Task	Total	Maximum	Average
1	3	2	1
2	3	2	1
3	2	1	2/3
4	1	1	1/3

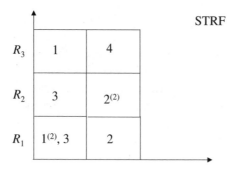

Fig. 3.5. Schedule obtained by means of STRF

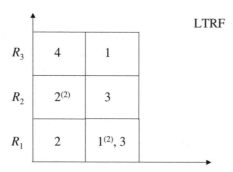

Fig. 3.6. Schedule obtained by means of LTRF

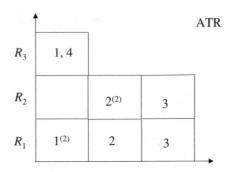

Fig. 3.7. Schedule obtained by means of ATR

3.4 The Metaheuristic Approach

In general, when the schedule makespan has to be minimized, techniques which tend to avoid solution *similarities* are designed. Indeed, in our problem the attempt to find a minimum value for *dev* and *peak*, requires that the number of time slots be not decreased rapidly. On the contrary, it seems worthwhile to analyze schedules with the same length before finding a lower makespan.

For the above reason, the function devoted to keeping the makespan low is a greedy one and is denoted as GS (Greedy Scheduler). GS is the starting function of the algorithm; it works greedily, considering tasks by non-increasing priorities, and allocates them one by one to the lowest available time slot without generating conflicts. At the beginning, before GS starts, each task $T_i \in \mathcal{T}$ is assigned a priority p_i according to its incompatibilities, *i.e.*, tasks with a higher potential for conflicts are tentatively scheduled first (ties are broken arbitrarily). Once GS has run, let t_i be the time slot tentatively assigned to task i, $T = max_{i \in \mathcal{T}}\{t_i\}$ be the total number of time slots used, and dev_{GS} be the unbalance level associated to the current schedule found by GS.

After this step, we try to improve *dev* without increasing the length T of the schedule found by GS. To accomplish this, we designed a function, denoted as DD, *i.e.*, *Dev Decreaser*, which attempts to move one or more tasks out of their current time slots. As with GS, before DD starts, each task is assigned a priority according to which it is visited by DD. This is done as follows. Consider time slot τ corresponding to the largest $tc_k^{(\tau)}$, with $k = 1, \ldots, K$. Tasks scheduled at τ are ordered by a non-decreasing value of r_{ik}, with $k = 1, \ldots, K$, and ties are broken in favor of tasks with the next highest r_{ik}. When tasks scheduled at τ are ordered, the next time slot τ' with the highest $tc_k^{(\tau')}$ is considered and the ordering process progresses until all the tasks are assigned a place.

According to the order found, DD schedules task i in the first feasible time slot (not exceeding T) where this new assignment will decrease *dev* (details on the neighborhood definition are given in the next subsection). If DD is not successful, *i.e.*, it is not able to change the schedule and thus $dev_{GS} = dev_{DD}$, a new function, denoted as PT (*Peak* Trader), is invoked (its functionalities will be explained later on). Otherwise, *i.e.*, $dev_{DD} < dev_{GS}$, DD starts its execution again and iterates until no further dev reduction can be obtained; at this point each task $T_i \in \mathcal{T}$ is assigned a priority $p_i = t_i$, and GS is restarted (ties are broken arbitrarily). We remark that, in this attempt to decrease dev, DD could find a schedule with even fewer time slots, *i.e.*, moving tasks from time slots to other time slots could return a new schedule with an empty time slot. If this happens, the algorithm compacts the schedule to $T - 1$ time slots and computes dev accordingly.

When dev_{GS} cannot be changed by DD, we get stuck as we are no longer able either to improve dev of the last schedule found by GS or to improve its length since a further invocation to GS would generate the same schedule. Thus, as stated above, a new function denoted PT is invoked with the goal of trading off *peak* for time slots. In other words, we check whether *peak* of the current schedule can be decreased by allocating an additional time slot (*i.e.*, $T+1$). To do this, a subset of tasks should be moved from its current position to $(T+1)$, and this choice is done as follows. Define for each task $T_i \in \mathcal{T}$ the quantity Δ_i as the difference between the current *peak* and *peak* associated with the same schedule where task i is moved to slot $T+1$. Once each task T_i is assigned a value Δ_i, PT neglects tasks with $\Delta_i < 0$ and selects the task with the maximum value of Δ_i which is then assigned to time slot $T+1$. Now, Δ_i values are recalculated and the next task i associated with the maximum nonnegative Δ_i and compatible with the tasks already assigned at $T + 1$, is assigned to time slot $T + 1$. At the end of this process, the priorities are reassigned as $p_i = 1/t_i$ and GS is invoked again (ties are broken arbitrarily). Note that although tasks with $\Delta_i = 0$ give no *peak* improvement in the solution, they are considered by PT as they can be useful for the generation of similar solutions, and thus, may find possibly lower values of dev.

The algorithm stops when, after a certain number of iterations, no improvement is obtained. In order to schematize the algorithm functionalities described above we have sketched its steps in Table 3.4.

3.4.1 Conceptual Comparison with Known Local Search Methods

In order to put our local search into perspective with other existing methods, it should be noted that (see also Table 3.4) the iterative calls to GS allow the proposed algorithm to be interpreted as a *multi start greedy heuristic*. Further pursuing this connection, the presence of DD and PT, which confers the characteristics of a local search to our algorithm, renders our iterative multi start scheme similar to a *Greedy Random Adaptive Search Procedure* (GRASP), *e.g.*,

Table 3.4. The local search algorithm

---

1. Assign priorities and execute GS;

   let dev_{GS} be dev associated with the schedule found.

2. Let $dev_{DD}^{old} = \infty$.

3. Assign priorities and execute DD;

   let dev_{DD}^{new} be dev associated with the schedule found.

4. If $dev_{GS} = dev_{DD}^{new}$ go to Step 6.

5. If $dev_{DD}^{new} = dev_{DD}^{old}$ then

   assign priorities and go to Step 1

else

   assign priorities and go to Step 3.

6. Assign priorities and execute PT; go to Step 1.

---

see [149]. In fact, a GRASP is formed of two phases: a construction phase (carried out by a greedy algorithm) and a local search phase. In our algorithm, the construction phase is performed by GS and, as mentioned above, the local search operators are represented by DD and PT. As can be inferred by the definition of a GRASP, there is a main difference between the two approaches, *i.e.*, the greedy construction phase in a GRASP is randomized, while in our algorithm it is deterministically evaluated. Moreover, a GRASP at a generic stage, of the construction phase, randomly chooses the next element to be placed in the solution from a subset of the not yet allocated elements, the so called *Restricted Candidate List* (RCL) [90], which leads to the definition of a semi-greedy heuristic.

There is a further important consideration on DD and PT. Indeed, we note that, given a generic greedy solution, either the former or the latter performs a local search on such a schedule. This sheds light on the hybrid behavior of the proposed algorithm: when DD is active in decreasing the penalty, our approach behaves like a constructive multi-start greedy heuristic, whereas, it behaves as a sort of *strategic oscillation* method when PT is invoked [29, 75, 111]. In fact, it is clear from the description that PT operates like a destructive phase, as opposed to GS which acts as a constructive one.

In order to provide a deeper analysis, let us examine how the neighbor solutions are defined in our local search. By the definition of DD or PT, it is easy to see that we use a simple neighborhood generation. In fact, neighbor solutions are generated by the current solution by moving a single task to different feasible time slots. The neighborhood search is implemented using a first-improving strategy, *i.e.*, the current solution moves to the first neighbor whose dev is lower than that of the current solution. In our experiments, we also implemented the best improving strategy, *i.e.*, all neighbors are investigated and the current solution is replaced by the best neighbor. In practice, however, the two strategies arrived at the same solutions, while the strategy we adopted was faster.

Applying the neighborhood definition DD can be interpreted as a sort of *hill climbing* algorithm. In fact, hill climbing starts with a solution, x, with score $s(x)$, and processes each task in some a priori determined order, determined a priori. For each task e, in this order, the algorithm considers every neighbor of x in which e is allocated a different time slot with respect to the one currently assigned. For all such feasible neighbors, the score is calculated and the neighbor x' with the minimum score is selected, and if $s(x) < s(x')$, the old solution is replaced by the new one. Subsequently, the algorithm moves to the next task in the given order. Note that, as for hill climbing, DD restarts regardless of whether the last iteration has successfully decreased dev or not.

This allows us to define the proposed algorithm as a multi-start greedy heuristic where a local search is performed by hill climbing and a sort of strategic oscillation is used as an additional diversification strategy when hill climbing is no longer able to improve the solution as depicted in Figure 3.8.

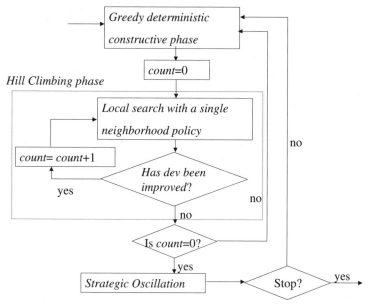

Fig. 3.8. The multi-start greedy algorithm structure. Parameter *count* is used to check if the local search in the hill climbing phase is executed more than once

3.4.2 How to Control the Effect of the Minimization of the Makespan and the Frequency Based Memory

It should be noted that the proposed approach is designed to allow the generation of similar solutions to balance resource allocation effectively which, however, can also have some drawbacks, as the algorithm can get stuck for a long time in particular areas of the solution space. To avoid this, we use a simple, but effective, *checkpointing* scheme [48]: the algorithm stops after a certain number of steps and then starts a new local search. In detail, define an *iteration* as an attempt (not necessarily successful) to change the time slot t_i of a task i. Roughly speaking, we force a checkpoint after a certain number of iterations: after each checkpoint, our algorithm starts another new local search from the current schedule. The number of iterations at which checkpointing is performed is constant, *i.e.*, checkpointing is performed exactly every ℓ iterations, with ℓ being a properly chosen constant.

After checkpointing, we introduce a careful assignment of priorities, which we call *bridging priorities*. The integration of these two features, *i.e.*, checkpointing and bridging priorities, allows one to start successive local searches which interact closely. We now define how the priorities are reassigned. Define a *phase* as the interval between any two successive checkpoints. Define the *update count* of a task i as the number of times i has changed its state in the current phase (*i.e.*, since the last checkpoint). Roughly speaking, the update count of a task measures its active participation during that phase. After a checkpoint, and before the update counts are reset, the task priorities are set as $p_i = 1/(update\_count + 1)$. With this assignment, a task with a low update count (*i.e.*, not very active in the last phase) gets a high priority, and thus there is a better chance of getting this task involved in the next phases. This clearly prevents the algorithm from continuing to visit the same subset of tasks. The checkpoint scheme presented allows one to correctly tune the range of different schedule lengths visited during the algorithm iterations. Forcing the algorithm to perform a small number of phases means that GS is invoked a small number of times, and thus we risk allowing the algorithm to visit schedules with high makespans, or similarly risk that the makespan expands progressively. On the contrary, if one forces the algorithm to perform a large number of phases, it invokes GS a large number of times, which means looking for lower makespans, or at least not increasing the current makespan. However, a good choice for the number of times GS has to be invoked (which is strictly dependent on the size of a phase) is very important in diversifying the solutions.

In Figure 3.9, we have plotted the values (identified by +) of the minimum makespan and the values of the maximum makespan (indicated by *) obtained by tuning the checkpoint to α times the number of tasks, *i.e.*, $\ell = \alpha \cdot n$ (the ex-

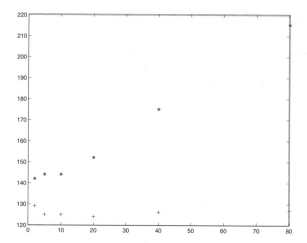

Fig. 3.9. The experiments are averages of 10 instances. The x-axis reports the value α; + are the lowest values of the makespan and * the highest ones

periments refer to the case of 500 tasks, 7 different resource types, a resource request of at most 10 units of each resource type and a density of incompatibilities equal to 0.8). The choice of these parameters will be discussed in Section 3.5.

Moreover, the way the bridging priorities are defined can be interpreted as the implementation of a frequency based memory. In fact, as for other iterated multi-start heuristics, one possible shortcoming of our heuristic without the checkpoint scheme could be the independence of its iterations, *i.e.*, the fact that it does not learn from the history of solutions found in the previous iterations. Information gathered from good solutions can be used to implement memory-based procedures.

In particular, the proposed frequency based memory is a sort of *principle of persistent attractiveness* (PPA) presented in [75]. The latter approach embodies an adaptive memory process by drawing on combinations of recency and frequency information, which can be monitored to encompass varying ranges of the search history. PPA says that good choices are derived from making decisions that have not previously been made during a particular phase of search even though that appeared to be attractive. That is, persistent attractiveness also carries with it the connotation of being "persistent unselected" during a specified interval. We have exploited this principle by creating a measure of attractiveness given by the *bridging priorities*.

3.5 Experimental Results

In this section, we discuss the experimental performance of our local search algorithm and make a comparison with lower bounds and results in the open literature. Our algorithm was implemented in the C language and run on a PC with a Pentium processor running at 500 MHz and 64 MB RAM. In what follows, we first describe the set up of the experiments which are presented and analyzed in Subsection 3.5.2. Finally, in Subsection 3.5.3 we compare them with the lower bounds discussed in Section 3.2, and in Subsection 3.5.4, we compare that with the algorithms presented in [138].

3.5.1 Description of the Experiments

Test instances have been generated randomly according to the following four parameters:

- n: the number of tasks;
- δ: the density of the *incompatibility graph* obtained as follows: there is a vertex in the graph for each task, and there is an edge between vertex i and vertex j if the corresponding tasks are incompatible and cannot be scheduled simultaneously;
- K: the maximum number of resource types;
- Q_{max}: the maximum amount of each resource type.

We let the parameter n assume the values 100, 250 and 500. δ is the edge probability in the incompatibility graph and ranges from 20% to 80% in increments of 20% (in the tables, we have used the notation 2 for 20%, 4 for 40%, and so on). The values of K used are 4 and 7, while Q_{max} is fixed at 10 for each resource type.

The parameters associated with the algorithm are:

- It_{max}: the number of iterations performed;
- $Check$: the number of iterations at which the algorithm starts a new phase (the checkpoint).

It_{max} in our experiments was fixed at one thousand times the number of tasks, while $Check$ was set equal to 5 times and 10 times the number of tasks.

For the sake of presentation, we divide the parameters characterizing the output into three sets (see Tables 3.6–3.9 in the next subsection). The first set of values is representative of our findings on the minimum makespan. The variables measured are:

- Mts: the minimum number of time slots found by the algorithm; ties are broken in favor of the schedule with the smallest dev;

- dev_{mts}: the value dev associated with the schedule of length Mts; ties are broken in favor of the schedule with the smallest $peak$;
- It_{mts}: the number of iterations at which Mts was achieved;
- CPU_{mts}: the running time, in seconds, at which Mts occurred;
- $peak_{mts}$: the peak of resource usage associated with the schedule of length Mts.

The second set of values, characterizing the output, refers to the minimization of dev, and involves:

- $Mdev$: the minimum value of dev found, and thus, the minimum deviation from a perfect workload balance; ties are broken in favor of the schedule with the smallest length;
- ts_{mdev}: the length of the schedule associated with the minimum value of dev; ties are broken in favor of the schedule with the smallest $peak$;
- It_{mdev}: the number of iterations at which $Mdev$ was achieved;
- CPU_{mdev}: the running time, in seconds, at which $Mdev$ occurred;
- $peak_{mdev}$: the peak of resource usage in this schedule.

The third and last set of values is associated with the best (minimum) peak of resource usage, and is represented by the following parameters:

- $Mpeak$: the minimum peak observed during the algorithm run; ties are broken in favor of the schedule with the smallest dev;
- dev_{mpeak}: the value of dev in the schedule associated with $Mpeak$; ties are broken in favor of the schedule with the smallest length;
- ts_{mpeak}: the length of such a schedule;
- It_{mpeak}: the number of iterations at which $Mpeak$ was achieved;
- CPU_{mpeak}: the running time, in seconds, the algorithm took to reach $Mpeak$.

Finally, in all the tables, we show how many seconds the algorithm took to terminate in row CPU.

In the next subsection, we give the experimental results on the 12 scenarios presented in Table 3.5.

3.5.2 Analysis of the Results

Let us now discuss the results in Tables 3.6–3.9. The most remarkable fact is that as long as δ grows, Mts grows together with ts_{mdev} and ts_{mpeak}. This can be explained by the fact that the number of incompatibilities increases with δ, and tasks are less easy to be scheduled in parallel. Moreover, this produces a decrease of $Mpeak$ which is strictly related to the density. Similarly, also $peak_{mts}$ and $peak_{mdev}$ decrease.

Table 3.5. The 12 scenarios examined

A_1 $K = 4$, $Q_{max} = 10$, $n = 100$, $It_{max} = 1,000 \cdot n$, $Check = 5 \cdot n$
A_2 The same as A_1 but $Check = 10 \cdot n$
B_1 $K = 7$, $Q_{max} = 10$, $n = 100$, $It_{max} = 1,000 \cdot n$, $Check = 5 \cdot n$
B_2 The same as B_1 but $Check = 10 \cdot n$
C_1 $K = 4$, $Q_{max} = 10$, $n = 250$, $It_{max} = 1,000 \cdot n$, $Check = 5 \cdot n$
C_2 The same as C_1 but $Check = 10 \cdot n$
D_1 $K = 7$, $Q_{max} = 10$, $n = 250$, $It_{max} = 1,000 \cdot n$, $Check = 5 \cdot n$
D_2 The same as D_1 but $Check = 10 \cdot n$
E_1 $K = 4$, $Q_{max} = 10$, $n = 500$, $It_{max} = 1,000 \cdot n$, $Check = 5 \cdot n$
E_2 The same as E_1 but $Check = 10 \cdot n$
F_1 $K = 7$, $Q_{max} = 10$, $n = 500$, $It_{max} = 1,000 \cdot n$, $Check = 5 \cdot n$
F_2 The same as F_1 but $Check = 10 \cdot n$

Comparing the results obtained for different values of $Check$ ($5 \cdot n$ and $10 \cdot n$) for the same scenario, when Mts values associated with $Check = 5 \cdot n$ are smaller than those obtained with $Check = 10 \cdot n$, the values of $peak_{mts}$ corresponding to $Check = 5 \cdot n$ are always greater than the one corresponding to $Check = 10 \cdot n$. This is justified by the trade-off existing between $peak$ and the makespan.

Comparing the results obtained for different values of $Check$ ($5 \cdot n$ and $10 \cdot n$) for the same class of instances, Mts values associated with $Check = 5 \cdot n$ are often lower than those corresponding to $Check = 10 \cdot n$. This result of the algorithm can be explained by the fact that a smaller value for $Check$ implies a higher number of calls to GS than when there is a bigger value for check. Thus, this provides more opportunities to reduce the makespan. In particular, in 58% of the instances tested, the algorithm with $Check = 5 \cdot n$, found a Mts smaller than the one found by fixing $Check = 10 \cdot n$, while in all the other cases the Mts values were equal.

Moreover, in 75% of the instances, using $Check = 10 \cdot n$ allows the algorithm to find Mts in a smaller number of iterations with respect to $Check = 5 \cdot n$ (compare the values of It_{mts}), and obviously, when this happens, it is also faster. However, has stated above, Mts values obtained with $Check = 5 \cdot n$ are often smaller than those obtained with $Check = 10 \cdot n$.

Note that in 26 instances out of 48, $Mpeak$ is associated with a $dev_{mpeak} = 0$. Furthermore, in 58% of the instances, it can be observed that $Mpeak$ obtained with $Check = 5 \cdot n$ is higher than the one achieved with $Check = 10 \cdot n$. This is justified by the effect of the checkpoint which produces more calls to PT and DD when $Check$ is bigger.

Table 3.6. Experimental results for $\delta = 2$

	A_1	A_2	B_1	B_2	C_1	C_2	D_1	D_2	E_1	E_2	F_1	F_1
Mts	9	9	9	9	17	18	17	18	29	30	29	30
dev_{mts}	0.00	1.50	5.14	8.57	2.00	0.00	16.00	4.86	7.50	1.50	40.86	3.43
$It_{mts} * 10^3$	44	309	687	79	94.75	19.75	39.75	42.25	139	114.5	339.5	192
CPU_{mts}	1	2	6	2	39	7	16	24	260	127	692	317
$peak_{mts}$	54	45	48	51	70	68	75	73	97	76	100	84
$Mdev$	0.00	0.00	1.71	2.00	0.00	0.00	0.00	0.00	0.00	0.00	0.00	0.00
ts_{mdev}	9	10	10	11	18	18	19	20	30	31	31	33
$It_{mdev} * 10^3$	44	54	549	453	285	19.75	222.25	56	41.5	12	258.5	132
CPU_{mdev}	1	0	1	7	12	7	90	30	55	11	535	219
$peak_{mdev}$	54	49	51	48	66	68	62	71	79	81	82	82
$Mpeak$	46	45	46	43	66	68	62	69	79	76	79	77
dev_{mpeak}	8.00	1.50	1.71	3.43	0.00	0.00	0.00	1.71	0.00	1.50	1.71	1.71
ts_{mpeak}	9	10	11	11	18	18	19	19	30	30	32	31
$It_{mpeak} * 10^3$	33	309	66	169	285	197.5	222.25	41	41.5	114.5	47	91.5
CPU_{mpeak}	0	2	5	2	12	7	90	23	55	127	105	154
CPU	10	12	9	11	101	138	110	160	669	620	917	847

Finally, we conclude this subsection by providing a further analysis of the performance of our algorithm in Figure 3.10.

3.5.3 Lower Bounds Comparison

In order to compare the results obtained in the previous subsection, we have implemented the lower bound LB_{peak} presented in Section 3.2. In Table 3.10, we have shown the % gap between $Mpeak$ and LB_{peak}. Reading over the table, it can be observed that what we have predicted in Section 3.2 is true. In fact, as long as the density grows, the gap between $Mpeak$ and LB_{peak} becomes bigger, because LB_{peak} is not able to take into account the effects of the incompatibilities. Moreover, as long as the number of tasks grows LB_{peak} becomes less effective since it is computed by considering $x_{it} \in [0, 1]$, with $i = 1, \ldots, n$ and $t = 1, \ldots, T$, (see again Section 3.2). Gaps range from 0.28 to 0.40 when $\delta = 2$ and from 0.52 to 0.62 when $\delta = 8$.

In order to show that the latter effect can be reduced and thus, bigger gaps are not caused by the poor performance of our algorithm, we have also implemented an exact algorithm to solve the IPMS and have experimented it on the same instances as those presented in the previous subsection. In

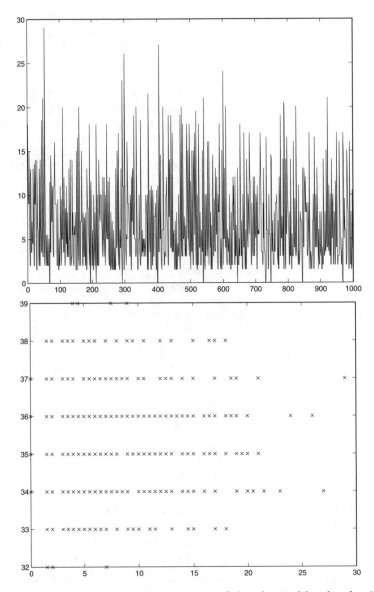

Fig. 3.10. Up: the chart shows the sequence of *dev* obtained by the algorithm in the scenario B_1 and $\delta = 8$. In the *x*-axis are the number of progressive invocation of the algorithm functions. Down: the chart shows the different *dev*-makespan combinations explored by the algorithm in 12,500 iterations on the same scenario B_1 with $\delta = 8$

Table 3.7. Experimental results for $\delta = 4$

	A_1	A_2	B_1	B_2	C_1	C_2	D_1	D_2	E_1	E_2	F_1	F_1
Mts	15	15	15	15	31	31	30	32	54	55	54	55
dev_{mts}	1.50	3.50	8.00	9.43	1.50	3.00	11.14	3.43	1.50	0.00	4.29	1.71
$It_{mts} * 10^3$	91.9	55.6	25.7	89.4	37.25	51	219.75	202.25	74	1.5	389.5	439.5
CPU_{mts}	11	6	5	12	24	38	131	146	101	1	737	1,003
$peak_{mts}$	36	43	38	37	52	52	51	46	55	55	51	49
$Mdev$	0.00	0.00	1.71	1.71	0.00	0.00	0.00	0.00	0.00	0.00	1.71	0.00
ts_{mdev}	16	17	18	16	32	33	35	34	55	55	57	58
$It_{mdev} * 10^3$	72.8	29.2	68.8	34.9	40.75	13.5	6.75	38.25	1.5	1.5	41.5	202
CPU_{mdev}	9	4	12	4	25	9	3	24	1	1	73	456
$peak_{mdev}$	38	37	31	31	47	46	42	46	55	55	51	48
$Mpeak$	30	32	31	31	43	46	42	45	54	55	50	48
dev_{mpeak}	1.50	1.50	1.71	1.71	0.00	0.00	0.00	2.00	3.00	0.00	2.86	0.00
ts_{mpeak}	17	16	18	16	34	33	35	35	54	55	59	58
$It_{mpeak} * 10^3$	13	183	688	349	135	135	67.5	11	73.5	1.5	26.5	202
CPU_{mpeak}	0	3	12	4	7	9	3	8	101	1	46	456
CPU	12	10	16	14	139	147	152	172	913	919	970	1,137

particular, we have implemented the column generation algorithm presented in [171]. Other exact algorithms can be found, *e.g.*, in [136, 137].

We ran the algorithm K times, associating each one to a single resource type scenario, and then took the maximum over the K runs (similarly to the way we defined LB_{peak}). Note that we experimented only with A_1, A_2, B_1 and B_2, since instances with 250 and 500 tasks are not likely to be solved exactly. In Table 3.11, we present the comparison of these results (given in column LB'_{peak}) with $Mpeak$.

In the experiments, the number of incompatibilities do not directly effect the quality of LB'_{peak}. Indeed, in IPMS $\mathcal{T}$ is supposed to be an independent set. Thus, the fact that LB'_{peak} varies in a given row is due to the fact that higher densities correspond to a higher number of time slots (see Tables 3.6–3.9), and the latter number determines the number of machines onto which tasks in $\mathcal{T}$ must be assigned for their execution. Trivially, in our tests $T = ts_{mpeak}$. The new gaps range from 0.02 to 0.09 when $\delta = 2$ and from 0.25 to 0.35 when $\delta = 8$.

Moreover, in order to compare Mts values with a lower bound, we have also optimally colored incompatibility graphs with $n = 100$ and $\delta = 2, 4, 6$ and 8 by means of the CDColor algorithm [47]. Note that we have not considered

Table 3.8. Experimental results for $\delta = 6$

	A_1	A_2	B_1	B_2	C_1	C_2	D_1	D_2	E_1	E_2	F_1	F_1
Mts	22	22	22	22	46	48	46	48	83	85	83	84
dev_{mts}	2.00	3.00	5.43	8.29	5.00	1.50	11.71	5.43	0.00	5.00	11.43	12.29
$It_{mts}*10^3$	71.8	60.7	94.9	33.4	107.25	21.75	247.25	67.25	354	158.5	314	109.5
CPU_{mts}	8	9	19	6	72	18	154	55	772	450	837	360
$peak_{mts}$	28	29	30	30	40	34	41	36	42	40	45	44
$Mdev$	0.00	0.00	0.00	0.00	0.00	0.00	1.71	0.00	0.00	0.00	0.00	1.71
ts_{mdev}	24	26	26	26	48	51	49	51	83	86	91	85
$It_{mdev}*10^3$	0.6	22.4	56.7	1.4	222	2	188.75	199.75	354	1.5	429	54.5
CPU_{mdev}	0	3	11	0	149	1	118	181	772	2	1,275	181
$peak_{mdev}$	25	26	24	24	33	33	49	35	42	38	39	39
$Mpeak$	25	26	24	24	33	33	34	32	37	38	39	38
dev_{mpeak}	0.00	0.00	0.00	0.00	0.00	0.00	2.00	1.71	1.50	0.00	0.00	1.71
ts_{mpeak}	24	26	26	26	48	51	47	53	84	86	91	90
$It_{mpeak}*10^3$	0.6	22.4	56.7	1.4	222	2	99.75	126	39.5	1.5	429	4.5
CPU_{mpeak}	0	3	11	0	149	1	64	104	100	2	1,275	17
CPU	13	12	20	15	168	203	157	250	941	989	1,303	1,649

larger graphs, *i.e.*, with $n = 250$ and $n = 500$, since (as for IPMS) exact solutions are very unlikely to be found on such graphs due to the complexity of vertex coloring [73]. Table 3.12 shows the % gap between Mts and the chromatic number of the incompatibility graphs (denoted as $Opt.$) associated with A_1, A_2, B_1 and B_2 (which have $n = 100$). The gaps range from 10% to 22%.

Finally, we ask the reader to note that no lower bound comparison has been made with respect to $Mdev$, since it differed from zero in only 8 cases out of the 48 tested, which is an index of its good performance.

3.5.4 Comparison with Known Algorithms

In this subsection, we compare the performance of our algorithm with those of the algorithms presented in [138], considering the objective of minimizing *peak*. In [138], the authors present four heuristics, namely H_1, H_2, H_3 and H_4, each based on a different priority rule with which tasks are selected to be scheduled, *i.e.*, minimum parallelity (MPA), minimum slack time (MST), latest start time (LST), greatest resource demand (GRD) and strict order (SOR). In particular, H_1 chooses activities according to SOR and breaks ties

Table 3.9. Experimental results for $\delta = 8$

	A_1	A_2	B_1	B_2	C_1	C_2	D_1	D_2	E_1	E_2	F_1	F_1
Mts	31	32	32	32	69	71	69	71	124	127	125	129
dev_{mts}	6.00	0.00	7.14	10.86	1.50	3.00	2.00	3.71	3.50	1.50	2.00	2.86
$It_{mts} * 10^3$	75.9	26.4	92.8	84.4	64.25	91	246.75	188.25	424	241	329.5	82
CPU_{mts}	13	5	18	18	60	158	271	399	1,434	1,125	1,042	391
$peak_{mts}$	25	25	24	29	27	31	30	29	34	32	30	29
$Mdev$	0.00	0.00	1.71	0.00	0.00	0.00	0.00	0.00	0.00	0.00	0.00	0.00
ts_{mdev}	34	36	35	36	72	72	76	76	132	131	137	135
$It_{mdev} * 10^3$	11.3	8.3	6.5	99.8	1	1	99.25	24.75	71	314	232.5	136.5
CPU_{mdev}	1	1	1	23	1	1	117	53	209	1,413	759	659
$peak_{mdev}$	20	23	21	21	28	28	27	25	30	30	29	28
$Mpeak$	20	23	21	21	26	28	27	25	29	29	29	28
dev_{mpeak}	0.00	0.00	1.71	0.00	1.50	0.00	0.00	0.00	2.00	0.00	0.00	0.00
ts_{mpeak}	34	36	35	36	70	72	76	76	129	133	137	135
$It_{mpeak} * 10^3$	11.3	8.3	6.5	99.8	31.75	1	99.25	24.75	1	236.5	232.5	136.5
CPU_{mpeak}	1	1	1	23	29	1	117	53	1	1,110	759	659
CPU	16	21	19	23	229	335	279	510	1,482	1,517	1,653	2,480

on the basis of LST, H_2 uses MST and MPA, H_3 uses GRD and MST, and, finally, H_4 uses MST and GRD. To allow a fair comparison, since the algorithms of Neumann and Zimmermann are based on precedence relationships and minimum and maximum time lags constraint, we used the same setting as the one described in Subsection 3.5.1, except for the fact that the graphs generated are directed and acyclic. Moreover, unitary processing times are considered, and for the competing algorithms, we set $T = ts_{mpeak}$.

Although we are working with directed graphs, our algorithm can be adapted by making a simple change: in fact, when a task is moved from a time slot to another slot, precedence constraints have to be obeyed rather than incompatibility constraints.

In Tables 3.13-3.14, we reported the experimental results, where column NZ shows the min-max range of the results obtained by H_1, H_2, H_3 and H_4, and column $Mpeak$ shows the results achieved by our algorithm. Referring to NZ, we show in parentheses which of the four algorithms has achieved the reported $peak$ value. For instance, $49(H_1)$ $54(H_2)$ means that the best algorithm was H_1, which achieved 49, and the worst was H_2, which achieved 54. From Tables 3.13-3.14, it should be noted that the performance results of our algorithm on these instances is very similar to that observed in the

Table 3.10. Comparison of $Mpeak$ with LB_{peak}

δ	2			4			6			8		
	$Mpeak$	LB_{peak}	%	$Mpeak$	LB_{peak}	%	$Mpeak$	LB_{peak}	%	$Mpeak$	LB_{peak}	%
A_1	46	33	0.28	30	18	0.40	25	13	0.48	20	9	0.55
A_2	45	30	0.30	32	19	0.41	26	12	0.54	23	9	0.61
B_1	46	31	0.32	31	17	0.45	24	13	0.46	21	10	0.52
B_2	43	31	0.28	31	19	0.39	24	13	0.46	21	10	0.52
C_1	66	44	0.33	43	24	0.44	33	17	0.48	26	12	0.54
C_2	68	44	0.35	46	24	0.39	33	16	0.51	28	11	0.61
D_1	62	43	0.31	42	24	0.43	34	18	0.47	27	11	0.59
D_2	69	43	0.38	45	24	0.47	32	16	0.50	25	11	0.56
E_1	79	50	0.37	54	28	0.48	37	18	0.51	29	12	0.59
E_2	76	50	0.34	55	27	0.51	38	18	0.52	29	12	0.59
F_1	79	47	0.40	50	25	0.50	39	17	0.31	29	11	0.62
F_2	77	48	0.38	48	26	0.46	38	17	0.29	28	11	0.61

Table 3.11. Comparison of $Mpeak$ with LB'_{peak}

δ	2			4			6			8		
	$Mpeak$	LB'_{peak}	%	$Mpeak$	LB'_{peak}	%	$Mpeak$	LB'_{peak}	%	$Mpeak$	LB'_{peak}	%
A_1	46	44	0.04	30	27	0.10	25	20	0.20	20	15	0.25
A_2	45	44	0.02	32	27	0.16	26	20	0.23	23	15	0.35
B_1	46	42	0.09	31	26	0.16	24	18	0.25	21	14	0.33
B_2	43	42	0.02	31	26	0.16	24	18	0.25	21	14	0.33

previous subsections. Moreover, its competitiveness with respect to H_1, H_2, H_3 and H_4 is significant.

A final remark is devoted to the running time needed by the algorithms to achieve the best solution, which ranges from 0 to 100 seconds for our algorithm, and from 0 to 40 seconds for the competing algorithms. This highlights how the improvements were obtained in a reasonable amount of additional time. Moreover, it should be noted (*e.g.*, compare the running times given above with those in Tables 3.6–3.9) how the introduction of the precedence relationships reduces the number of admissible solutions, especially when the density is very high, and how this impacts on the time needed to find the best solution.

Table 3.12. Comparison of *Mts* with *Opt*

δ	2			4			6			8		
	Mts	*Opt.*	%	*Mts*	*Opt.*	%	*Mts*	*Opt.*	%	*Mts*	*Opt.*	%
A_1	9	7	0.22	15	13	0.13	22	18	0.18	31	28	0.10
A_2	9	7	0.22	15	13	0.13	22	18	0.18	32	28	0.13
B_1	9	7	0.22	15	13	0.13	22	18	0.18	32	28	0.13
B_2	9	7	0.22	15	13	0.13	22	18	0.18	32	28	0.13

Table 3.13. Comparison between our algorithm and the algorithms in [138]

δ	2			4	
	Mpeak	*NZ*		*Mpeak*	*NZ*
A_1	48	$49(H_1)$	$54(H_2)$	32	$35(H_1)$ $38(H_3)$
A_2	48	$49(H_1)$	$54(H_2)$	33	$35(H_1)$ $38(H_3)$
B_1	47	$48(H_1)$	$53(H_2)$	32	$31(H_1)$ $34(H_2)$
B_2	45	$48(H_1)$	$53(H_2)$	31	$31(H_1)$ $34(H_2)$
C_1	69	$71(H_4)$	$75(H_3)$	45	$47(H_4)$ $48(H_2)$
C_2	72	$71(H_4)$	$75(H_3)$	48	$47(H_4)$ $48(H_2)$
D_1	65	$68(H_4)$	$74(H_2)$	43	$45(H_4)$ $48(H_2)$
D_2	72	$68(H_4)$	$74(H_2)$	46	$45(H_4)$ $48(H_2)$
E_1	83	$87(H_1)$	$92(H_3)$	55	$57(H_1)$ $62(H_3)$
E_2	80	$87(H_1)$	$92(H_3)$	56	$57(H_1)$ $62(H_3)$
F_1	82	$85(H_1)$	$89(H_2)$	51	$50(H_4)$ $54(H_2)$
F_2	80	$85(H_1)$	$89(H_2)$	49	$50(H_4)$ $54(H_2)$

3.6 The Extension to the Case with Arbitrary Integer Duration

When arbitrary weights are considered, the algorithm functions change to cope with assignments of set of a consecutive number time slots equal to the task duration, rather than single time slots. Let p_j be the integer duration of task j.

As described in the previous sections, function GS (Greedy Scheduler) greedily considers tasks according to non-increasing priorities. Assume that j is the next task to be allocated: GS assigns to j a set of time slots with cardinality equal to p_j such that each one of the time slot does not generate conflicts with the existing ones, and the time slots assigned are consecutive.

Table 3.14. Comparison between our algorithm and the algorithms in [138]

δ		6			8	
	$Mpeak$	NZ		$Mpeak$	NZ	
A_1	26	$32(H_1)$	$36(H_3)$	20	$22(H_1)$	$28(H_2)$
A_2	26	$32(H_1)$	$36(H_3)$	23	$22(H_1)$	$28(H_2)$
B_1	26	$32(H_1)$	$36(H_2)$	21	$25(H_1)$	$32(H_2)$
B_2	26	$32(H_1)$	$36(H_2)$	21	$25(H_1)$	$32(H_2)$
C_1	34	$34(H_4)$	$36(H_2)$	26	$28(H_4)$	$34(H_3)$
C_2	34	$34(H_4)$	$36(H_2)$	28	$28(H_4)$	$34(H_3)$
D_1	34	$34(H_1)$	$38(H_2)$	27	$30(H_1)$	$34(H_2)$
D_2	34	$34(H_1)$	$38(H_2)$	25	$30(H_1)$	$34(H_2)$
E_1	37	$40(H_1)$	$45(H_2)$	29	$30(H_1)$	$35(H_2)$
E_2	38	$40(H_1)$	$45(H_2)$	29	$30(H_1)$	$35(H_2)$
F_1	40	$42(H_4)$	$46(H_3)$	29	$30(H_4)$	$35(H_3)$
F_2	42	$42(H_4)$	$46(H_3)$	28	$30(H_4)$	$35(H_3)$

Once GS has run, let set_i be the set of time slots tentatively assigned to task i, and $T = max_{i \in \mathcal{T}} max_{t_i \in set_i}\{t_i\}$ be the total number of time slots used, and dev_{GS} be the imbalance level associated with the current schedule found by GS.

The algorithm now tries to decrease dev without increasing the length T of the schedule offered by GS, by means of DD, i.e., Dev Decreaser. Similarly to the case with unitary durations, DD attempts to move one or more tasks from their current set of time slots. As with GS, before DD starts, each task is assigned a priority with which it is visited by DD. This is done as follows. Consider time slot τ corresponding to the largest $tc_k^{(\tau)}$, with $k = 1, \ldots, K$. Tasks scheduled at τ are ordered by non decreasing value of r_{ik}, with $k = 1, \ldots, K$, and ties are broken in favor of tasks with the successive highest r_{ik}. When tasks scheduled at τ are ordered, the next time slot τ' with the highest $tc_k^{(\tau')}$ is considered and the ordering process progresses until all the tasks are assigned a place.

According to the order found, DD schedules task i in the first place in the schedule where p_i time slots are available (not exceeding T) when this new assignment decreases dev. If DD is not successful, i.e., it is not able to change the schedule and thus $dev_{GS} = dev_{DD}$, PT (Peak Trader). Otherwise, i.e., $dev_{DD} < dev_{GS}$, DD starts again its execution and iterates until no further dev reduction can be obtained; at this point each task $T_i \in \mathcal{T}$ is assigned a priority $p_i = max_{t_i \in set_i}\{t_i\}$, and GS is restarted (ties are broken arbitrarily).

When dev_{GS} cannot be changed by DD we get stuck as we are no longer able neither to improve dev of the last schedule found by GS nor to improve its length since a successive call to GS would generate the same schedule. Thus, PT is invoked with the goal of trading off $peak$ for time slots. As with the case of unitary task durations, the algorithm tries to decrease $peak$ of the current schedule by allocating some additional time slots. To do this, a subset of tasks SS should be moved from their current time slots to the piece of schedule from $(T + 1)$ to $\max_{i \in SS} \max_{t_i \in set_i} t_i$, and this choice is done as follows. Define for each task $T_i \in \mathcal{T}$ the quantity Δ_i as the difference between the current $peak$ and $peak$ associated with the same schedule where task i is moved over T. Once each task T_i is assigned a value Δ_i, PT neglects tasks with $\Delta_i < 0$ and selects the task with the maximum value of Δ_i which is then assigned to time slots over T. Now, Δ_i are recalculated and the next task i associated with the maximum nonnegative Δ_i and compatible with tasks already assigned over T is assigned to the time slots over T. At the end of this process the priorities are reassigned as $p_i = 1/\max_{t_i \in set_i} t_i$ and GS is invoked again (ties are broken arbitrarily). As with the other algorithm, tasks with $\Delta_i = 0$ are considered by PT as they can be useful for the generation of similar solutions and thus for finding possibly lower values of dev.

As with the case with unitary duration, the algorithm stops when, after a certain number of iterations, no improvement is obtained; moreover, when facing directed graphs, the weighted case can be adapted extending the simple change presented in the previous section: when a task is moved from a set of time slots to another set of time slots, precedence constraints have to be obeyed rather than incompatibilities constraints.

3.7 Case Study 1

A department of a telecommunications company produces transmission and reception antennas. The department is in charge of producing three antenna types (a1, a2, a3). The production layout is depicted in Figure 3.11.

These are associated with the routings:

- a1 (solid line),
- a2 (dotted line),
- a3 (dashed line).

We describe the assembly process for each antenna (Figure 3.12 shows a picture of the process). There are 7 major tasks, as described in the following.

- Assembly initialization. In this macro operation the collection of parts, the cleaning of components, and the electrical cable connection take place. Moreover, the major components are assembled.

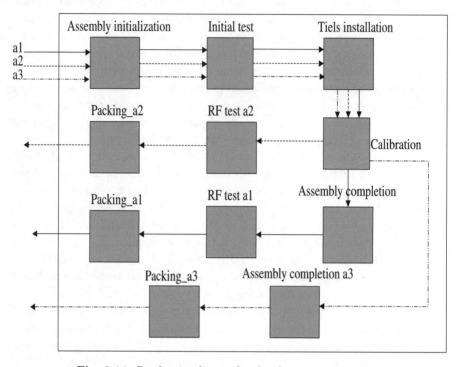

Fig. 3.11. Production layout for the three antenna types

- Initial Test. This operation consists of a switch on test performed by a robot sending electrical impulses. Troubleshooting is reported.

- Tiels installation. In this task the tiels (panels) for reception and transmission are mounted on the antenna by robots.

- Calibration. This is a second test process, executed inside a large electrical machine. A robot sends impulses to the antenna and by means of specific software, the output signal and its error are measured.

- Assembly completion. The main external panels are mounted on the antenna. This task is similar to task Assembly initialization, but it is shorter. Assembly completion is a task to be executed exclusively by antenna type 1.

- **Assembly completion a3.** The same as Assembly completion; it refers to antenna type 3.

- **RF test a1.** In this process, a test on the transmission and reception capacity of the antenna is performed. This requires a special shielded room where a probe robot emits electromagnetic fields simulating the existence of signal from different distances and angles. Label a1 indicates that the process is dedicated to antenna type 1.

- **RF test a2.** The same as RF test a2; it refers to antenna type 2.

- **Packing a1.** For each antenna, the packing operation requires a first phase of packing and a successive phase of stocking the finished products in the warehouse, where they remain until delivery. As for the RF test, label a1 indicates that the process is dedicated to antenna type 1.

- **Packing a2.** The same as Packing a1; it refers to antenna type 2.

- **Packing a3.** The same as Packing a1 and Packing a2; it refers to antenna type 3.

Table 3.15. Duration and resources associated with each task of the antenna assembly process. Task duration are in hours

Task	Antenna 1		Antenna 2		Antenna 3	
	Time	Technicians	Time	Technicians	Time	Technicians
Assembly initialitation	5	4	5	4	5	4
Initial test	3	2	3	2	3	2
Tiels intallation	1	2	1	2	1	2
Calibration	3	2	3	2	3	2
Assembly completion	3	2	-	-	-	-
Assembly completion a3	-	-	-	-	4	3
RF test a1	4	2	-	-	-	-
RF test a2	-	-	3	2	-	-
Packing a1	4	2	-	-	-	-
Packing a2	-	-	5	4	-	-
Packing a3	-	-	-	-	2	1

Fig. 3.12. The antenna assembly process

In Table 3.15 we give the task duration of each antenna type.

Considering the routings represented in Figure 3.11 we have the precedence relationships among tasks shown in Table 3.16.

Consider a new order arriving at the telecommunications company. The order requires the production of 15 different antennas for each antenna type. In the contract, a deadline of D days starting from the signature of the order is specified.

Moreover, if the delivery is made before D, an extra reward of Y thousand Euro will be provided for each day saved. As the company is a large one, the Production Manager (PM) is quite sure that the production plant has enough capacity for this new order. Although the plant is highly automated, specialized technicians are required during several parts of the process. The number of requested personnel is given in Table 3.15.

The Problem

To find out what the situation is like, the PM asks for a master schedule regarding this order with the following objectives:

Table 3.16. Precedence relations among tasks

Task	Predecessors
Assembly initialitation	-
Initial test	Assembly initialitation
Tiels intallation	Initial test
Calibration	Tiels intallation
Assembly completion	Calibration
Assembly completion a3	Calibration
RF test a1	Assembly completion
RF test a2	Calibration
Packing a1	RF test a1
Packing a2	RF test a2
Packing a3	Assembly completion a3

- Find the minimum time horizon for the delivery.

- Find the maximum number of specialized technicians.

For this purpose the PM models the problem adopting some simplifying assumptions:

- Some production equipment (machines, testing platforms, and others) have unit capacity and are required by more than one task. This implies that some tasks cannot be executed simultaneously, even though there is enough manpower.

- The number of available technicians for each task is known (see again Table 3.15).

- Duration are integers numbers.

The Graph Model Associated with the Problem

In Figure 3.13 we give the graph where arcs represent precedence constraints and edges (non oriented arcs) model resource incompatibilities. Table 3.17 shows the mapping between the node numbering of the graph and the tasks of the production process.

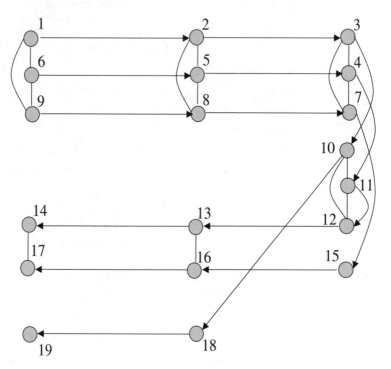

Fig. 3.13. The graph associated with the production process

The Solution

Applying the algorithm proposed with arbitrary weights, we obtain a solution of 36 hours for a production cycle, *i.e.*, 180 hours (15 days) to produce all the 15 triples of antennas, and a peak of 6 technicians. If we compare this solution with that of the greedy algorithm we obtain 45 hours per cycle and a total of 220 hours (19 days) with a peak of 5 technicians. Now, if D is 17 days and we adopt the metaheuristic solution, we obtain an advantage in terms of cost if the money earned for two day savings is bigger than the sum of the cost of an additional technician and the cost of two days of deadline violation.

3.8 Case Study 2

In this case study, we consider a small foundry. This company does not handle inventory systems and products are produced on demand for customers. The production line has six stages and there is an expert in charge of daily

Table 3.17. Duration and resources associated with each task of the antenna assembly process. Task duration are in hours

Task	Antenna 1 Node label	Antenna 2 Node label	Antenna 3 Node label
Assembly initialitation	6	9	1
Initial test	5	8	2
Tiels intallation	4	7	3
Calibration	11	15	10
Assembly completion	12	-	-
Assembly completion a3	-	-	18
RF test a1	13	-	-
RF test a2	-	16	-
Packing a1	14	-	-
Packing a2	-	17	-
Packing a3	-	-	19

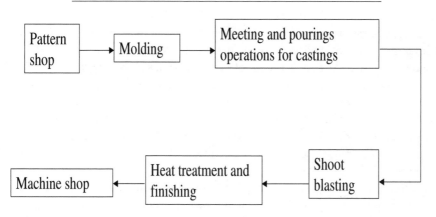

Fig. 3.14. The production line

production planning. The expert is a production chief and the decisions are made based solely on past experience.

In Figure 3.14 we show the production line. A product remains at a given stage for a variable time, depending on many factors (basically physical features), and operations that are to be accomplished at each stage depend on human and/or machine resources. The most crucial stage is Fusion (see Figure 3.14), which encompasses Molding and Shoot Blasting. This stage is always done on site. For other stages, it may be possible to outsource if an excessive workload occurs. An exception might be Heat Treatment, as some products require secondary heat treatment after Machining. Each product is associated with a manufacturing schedule which relies on two factors: the parameters of the product (type, size, weight, *etc.*) and the kind of alloy to be used. In Fusion and Heat Treatment, time is computed depending on product weight and the alloy to be considered. Purchase Orders (PO) are the core element of production planning. A PO contains a key, customer data, and product data as well as specifications on the necessary production processes, costs and delivery dates. Customers are linked to one purchase order for each required alloy, which is unrelated to product quantities. So each order could contain from one to n products, each product having a variable quantity of component parts. Planning is focused on obtaining the optimal yield of the critical resource (alloy) as well as completing the PO in the shortest possible time.

Planning, as defined above, exists to help in decision making. If, or when, a new order appears, the system is able to recalculate the overall delay in the production plan and/or for each specific order. Recalling the $\alpha|\beta|\gamma$ notation, see Chapter 1, the problem is defined as:

$$FFs|Set\ of\ constraints|C_{max}.$$

The environment is defined as a Flexible Flow Shop, because in each of the six stages there are some input resources (used simultaneously) for the job. The parameters are grouped into three sets: Stages, Resources and Products, where: $E = \{E_1, \ldots, E_m\}$ is the set of m production stages; $R = \{R_1, \ldots, R_m\}$ set of m resource types, each type having k resources; $P = \{P_1, \ldots, P_n\}$ set of n products considered for planning; $|E| = m$ is the cardinal of set E, $|R| = m$ is the cardinal of set R; $|P| = n$ is the cardinal of set P.

For each production stage E_i, there is a duration time T_{ij} that depends on product P_j. Production stages are sequentially sorted and no permutation is allowed. This means that stage E_{i+1} cannot be done before stage E_i. Each product P_j has one production stage, as a minimum, and m production stages as a maximum. Each production stage E_i corresponds to only one type of resource from R_i. The selected resource R_i belonging to type i needed to manufacture product P_j, is the one that is available or, if not available, is the one for which time availability is closest to the current time, in which case product P_j has to wait for resource R_i. For every type of resource R_i, the quantity of elements, k, may be different. The quantity n of resources is limited only by the company category. The resources may be people or equipment. Finally, the problem is obtaining high production line utilization as a method to manufacture products in the shortest possible time. It is known in the literature that the makespan for this problem is defined as:

$$C_{max} = \max\{t_f(X_{1n}), \ldots, t_f(X_{kn})\} - \min\{t_i(X_{11}), \ldots, t_i(X_{k1})\}$$

where

$$\max\{t_f(X_{1m}), \ldots, t_f(X_{nm})\},$$

applies on the ending times of the last productive stage m for the n products, and function

$$\min\{t_i(X_{11}), \ldots, t_i(X_{m1})\},$$

applies on the starting times of the first productive stage.

3.9 Conclusions

In this chapter, we have studied the problem of assessing resource usage in scheduling with incompatibilities. Contrary to what appears in the literature, we have proposed a local search algorithm which simultaneously takes into consideration three different objective functions, $i.e.$, the minimization of the resource imbalance, and the minimization of the peak of resource usage and the makespan . Extensive experiments on synthetic data, as well as a comparison with lower bounds and known heuristics have been presented. The latter tests have been performed by adapting our algorithm to incorporate precedence constraints. What has emerged is that the proposed local search is able to find high quality solutions, especially if we consider the difficulty of dealing simultaneously with the three problems considered. Two case studies are presented at the end of the chapter.

4

Scheduling Jobs in Robotized Cells with Multiple Shared Resources

A substantial amount of recent research has been directed toward the development of industrial robots. In this chapter, we study the problem of sequencing jobs that have to be executed in a flexible cell, where a robot is the bottleneck resource. Indeed, as opposed to flow shops where jobs are typically moved by means of conveyors or similar transport systems, in this problem the robot handles the main task of managing the operations of loading the job onto a machine, unloading it after it has been processed, and, when requested, holding the job during the processing itself.

In this kind of production process, effective sequencing procedures are sought to prevent the robot from producing excessive delays in the maximum completion time.

4.1 Background and Problem Definition

The bulk of this work has dealt with electromechanical skills, sensing devices and computer controls. Relatively little research has investigated the operational problem associated with the application of this technology [5, 11, 41, 86, 106]. We investigate an operational problem, which is encountered in applications, in which a robot is used to tend a number of machines. Such an application would arise, for example, when machines have been organized into a machine cell to implement the concept of group technology [1, 2, 3, 4, 13, 170]. The cell would be used to produce multiple sets of parts at prespecified production rates. The feasibility of assigning to one robot all the tasks needed to tend all the machines, so that parts are produced at specified production rates, is an important operational problem. In fact, its solution would determine the number of robots needed to tend machines in a manufacturing system and hence the investment required to robotize tending activities.

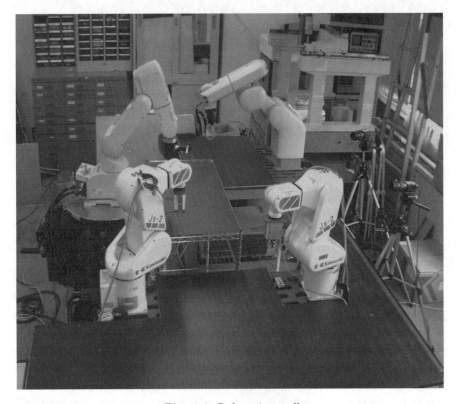

Fig. 4.1. Robots in a cell

This kind of problems was first introduced by Asfahl [12]. A summary of the related literature can be found in [156] and in [85]. In particular, in [156], the authors find the optimal sequence of moves for a two machine robot-centered cell producing a single part-type, and solve the part sequencing problem for a given one-unit robot move cycle in a two machine cell producing multiple part-types. In [85], Hall *et al.* showed that the optimal solution to the multiple part-types problem in a two machine cell is not generally given by a one-unit cycle. They developed an $O(n^4)$ time algorithm (n is the number of parts) that jointly optimizes the robot move cycle and part sequencing problems. The latter algorithm was later improved by Aneja and Kamoun in [11] with one of $O(n \log n)$ complexity.

Blazewicz *et al.* [39] also provided a summary of the related literature and described a production line for machine castings for truck differential assemblies in the form of a three-machine robotic cell where a robot has to move heavy mechanical parts among large machines.

Descriptions of interesting applications have been provided *e.g.* by Hartley [91]. Complexity issues were discussed by Wilhelm [175], and by Crama and Van De Klundert [60], that provided a proof that a basic version of the problem is strongly $\mathcal{NP}$-complete, and described a polynomial time algorithm for minimizing cycle time over all one-unit cycles in an m machine cell producing a single part-type in a robot-centered cell.

Hall *et al.*, in [86], considered a three machine cell producing multiple part-types, and proved that, in two out of the six potentially optimal robot move cycles for producing one unit, the recognition version of the part sequencing problem is unary $\mathcal{NP}$-complete. Moreover, they showed that the general part sequencing problem not restricted to any robot move cycle in a three machine cell is still intractable.

Levner *et al.*, in [122], addressed a cyclic robot scheduling problem in an automated manufacturing line in which a single robot is used to move parts from one workstation to another, with the goal of optimizing the cycle length. For this problem they proposed an algorithm of complexity $O(m^3 \log m)$, where m is the number of machines.

Many of the results and algorithms in the literature are devoted to robotic flow-shop scheduling problems (*e.g.*, see [12, 39, 59, 85]). However, the robotic cell configuration is very flexible: the robot can easily access the machines in any order, thus producing a large variety of products in the form of a job-shop (*e.g.*, see [74, 87, 98]).

In this chapter, we concentrate on the latter class of problems, studying the general problem of sequencing multiple jobs where each job consists of multiple ordered tasks and task execution requires the simultaneous usage of several resources [22].

The remainder of the chapter is organized as follows. In Section 4.2 we formally describe the problem. The complexity of the problem is analyzed in Section 4.3. A heuristic algorithm is described in Section 4.4, and finally, in Section 4.5 we present some computational results.

4.2 Problem Definition

Let us consider a cell composed of M machines configured to produce a batch of parts. Part p, $p = 1, ..., P$, requires N_p tasks. Let p_j denote the jth task for part p. Task p_j takes t_{p_j} time units to complete. We allow the general condition that a task requires the concurrent usage of more than one cell resource during its execution. Let S_{p_j} be the set of resources required for task p_j. The objective is to determine the schedule for all tasks of all parts so as to minimize the makespan. Note that the mainstream of research in robotic scheduling is devoted to two classes of production performance measures. The first is the makespan, which addresses the case of a finite part set where one is interested in minimizing the maximum completion time of these parts (*e.g.*, see [54, 109, 110]). The other class of models (which is not addressed in this chapter) assumes that the part set is infinite and attempts to minimize the long run cycle time of the schedule, which is the same as maximizing the throughput rate (*e.g.*, see [156]).

As an example of this model, we examine an assembly cell of $M - 1$ machines plus a material handling robot. The robot will be modelled as machine M. Each part has a processing time associated with each machine. In addition, the robot is needed to load a part onto a machine and then to unload the part after the production task has been performed. In our model, each production task is divided into three tasks, namely Load, Make, and Unload. The Load and Unload tasks require both the machine and the robot. The Make operation would normally require only the machine. The formulation also permits the case where the robot is needed to hold the part while certain production tasks are performed. This problem can be easily extended to the case of multiple material handlers, each assigned to a specific set of machines. Note that the solution must take into account precedence restrictions. For instance, we cannot unload until we have loaded and performed the Make task.

This is an example of an operational problem associated with the application of industrial robots used to tend a number of machines that have been organized into a machine cell. The cell is used to produce a set of parts at production rates specified by managers. The time required to tend a machine may be different for each machine according to its location and orientation in the cell, and to the part to be processed. All machines in the cell are dependent on a single robot, so the sequence in which tending tasks are performed may be critical and force certain machines to be idle.

Let us denote the Load, Make and Unload operations of part p requiring machine i as L_p^i, M_p^i, U_p^i (the machine index i will be omitted when not necessary). Moreover, we shall refer to the problem of scheduling Load, Make and Unload operations for the set of parts in order to minimize makespan as the LMU Problem.

As an example of the LMU Problem, in Figure 4.2 is shows a schedule for three parts and two making machines. In particular, the robot is indicated by R, and the machines by m_1 and m_2, respectively. Moreover, the Make operation for Part 2, namely M_2^2, requires resource R.

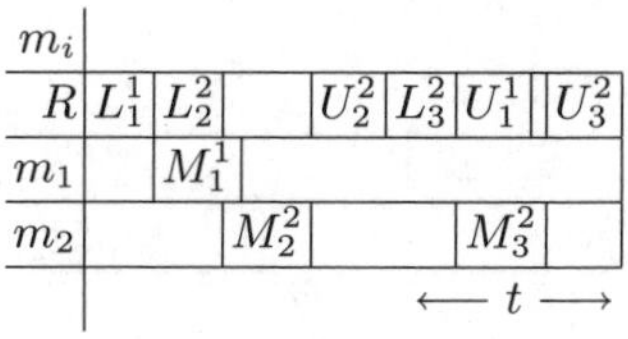

m_i								
R	L_1^1	L_2^2			U_2^2	L_3^2	U_1^1	U_3^2
m_1		M_1^1						
m_2			M_2^2			M_3^2		

$\longleftarrow t \longrightarrow$

Fig. 4.2. An example of the LMU Problem

4.3 $\mathcal{NP}$-Completeness Result

We show that the LMU Problem with general processing times on four machines is $\mathcal{NP}$-hard by transforming the 3-Partition problem (strongly $\mathcal{NP}$-complete [73]) to our problem. The 3-Partition problem is defined as follows.

The 3-Partition problem: Given a set $A = \{a_1, a_2, \ldots, a_{3z}\}$ of $3z$ integers such that $\sum_{i=1}^{3z} a_i = zB$ and $B/4 < a_i < B/2$ for $i = 1, \ldots, 3z$, can A be partitioned into z disjoint subsets, $A_1, A_2, \ldots, A_z$, such that $\sum_{a_i \in A_k} a_i = B$ for each $k = 1, 2, \ldots, z$?

Theorem 3. *The LMU problem with m=4 is strongly NP-complete*

Proof: For a given instance of the 3-Partition problem let us define a corresponding instance of our problem with $5z$ parts. Let there be $2z$ parts requiring machine 1, z parts requiring machine 2, z parts requiring machine 3, and z parts requiring machine 4. Recalling the definition of B in the statement of the 3-Partition problem, let the processing times be as follows:

- $L_p^1 = 1, \ M_p^1 = B, \ U_p^1 = 1, \ p = 1, \ldots, 2z$
- $L_p^2 = a_{p-2z}, \ M_p^2 = B - a_{p-2z} + 2, \ U_p^2 = a_{p-2z}, \ p = 2z+1, \ldots, 3z$
- $L_p^3 = a_{p-3z}, \ M_p^3 = B - a_{p-3z} + 2, \ U_p^3 = a_{p-3z}, \ p = 3z+1, \ldots, 4z$
- $L_p^4 = a_{p-4z}, \ M_p^4 = B - a_{p-4z} + 2, \ U_p^4 = a_{p-4z}, \ p = 4z+1, \ldots, 5z$

If the 3-Partition problem has a positive answer, then we can construct a schedule of length $Y = 2z(B+2)$ as shown in Figure 4.3.

We will now show that there is a positive answer to the 3-Partition problem if there exists a feasible schedule with a length less than or equal to Y. We observe that the value of the makespan to schedule, in any order, the $2z$ parts requiring machine 1, is $Y = 2z(B+2)$. This partial schedule has no idle times on machine 1 and has $2z$ idle times on the robot for all of length B. The total time required to perform the Load and Unload operations of the remaining $3z$ parts is $2zB$. Note that a feasible schedule for the remaining parts can be obtained scheduling the Load operations of parts on machines 2, 3, 4, in any order, on an idle time of the robot. The Make operation can start as soon as

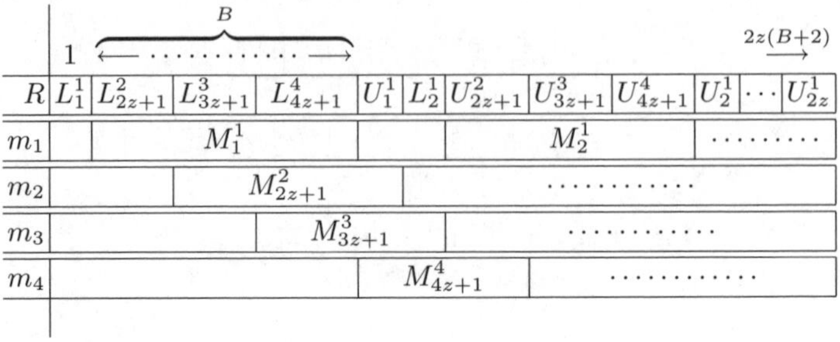

Fig. 4.3. An optimal schedule with $Y = 2z(B + 2)$

the Load is completed. Unload operations can be performed in the successive robot idle time in the same order as the corresponding Load operations. In a schedule of length Y the robot must have no idle times. This is possible if the sum of the Load (Unload) operations of parts requiring machines $2, 3$, and 4 in each robot idle time is equal to B, that is if a 3-Partition exists for set A. Thus the problem is strongly $\mathcal{NP}$-complete.

4.4 The Proposed Heuristic

In the following, we describe a heuristic algorithm (namely LMUA) which finds a feasible solution to our scheduling problem. First we observe that an active schedule can have idle times either on one of the machines or on the robot which cannot be eliminated trivially. The robot can be idle when all the machines are simultaneously processing a Make task. A machine m_i may be idle, waiting for either a Load or an Unload task to be performed, because the robot is busy tending another machine. Secondly, for any feasible schedule, the maximum completion time is the completion time of the last Unload operation.

The *LMUA* algorithm proposed is a single pass heuristic in which the loading-unloading sequence and the corresponding schedule are determined only once. A list containing the sequence of Load-Make-Unload tasks is built considering any order of the part types. At the beginning, the robot R loads all machines. Make operations can start immediately after the preceding Load is performed. Successively, the robot unloads the machine which ended the Make task first. In the generic stage of the algorithm the first unselected task in the list is examined. If it is a Make operation it can be scheduled immediately after the loading. Otherwise, the first Load-Unload operation in the list of remaining tasks which tends the machine that has been idle for the longest time is selected. The following is a pseudo-code of the *LMUA* algorithm.

The $LMUA$ Algorithm

Step 1. Consider an instance with $M-1$ machines, one robot (modelled as machine M) and Z parts.Take any ordering of all the parts (assume the order $1, \ldots, Z$);

Step 2. Build the list of tasks:

$$LT = \{L_1^k, M_1^k, U_1^k, L_2^k, M_2^k, \ldots, L_Z^k, M_Z^k, U_Z^k\};$$

build the list of Tasks Make that require the resource robot:

$$LTR = \{M_i^k| \text{ part } i \text{ requires the robot during the Make on } m_k\};$$

Build the list of processing times:

$$PT = \{p_{L_1^k}, p_{M_1^k}, p_{U_1^k}, \ldots, p_{U_Z^k}\};$$

build the list of the instants of time at which the machines are available:

$$AT = \{At_1, At_2, \ldots, At_M\};$$

Step 3. Initialize the current scheduling time at $t = 0$; initialize the list of the tasks that can be processed at the current t with all the Load tasks:

$$LTA = \{L_1^k, L_2^k, \ldots, L_Z^k\};$$

Step 4. Build the list reporting the instants of time at which the tasks in list LT can start their execution:

$$FTA = \{Ft_{M_1^k}, Ft_{U_1^k}, \ldots, Ft_{U_Z^k}\};$$

Step 5. Set the values of the variables in FTA equal to infinite;

Step 6. While $LT \neq \emptyset$ or $LTA \neq \emptyset$ do:

Step 6.1. Scan tasks in list LTA and if there exists a task Make that at time t requires a machine m_k that is available according to list AT, then go to Step 8; otherwise, if the robot is available at time t and there exists either a task Load whose corresponding Make operation requires a machine which is available, or a task Unload, then go to Step 12 (tie breaks choosing the Load or Unload task wait for more time in the list).

Step 7. If does not exist a task obeying Step 7.1.2 or Step 7.1.3, then Increase $t = t+1$; update lists LTA and LT by moving tasks from LT to LTA according to FTA.

Step 8. Schedule the Make task selected starting from t on the required machine;

Step 9. Set equal to infinite the variable in AT associated with the machine handling the Make task.

Step 10. Update in FTA the earliest starting time of the Unload task associated with the processed Make task, setting it to the finishing time of the latter;

Step 11. Delete from LTA the processed Make task. Set $t = t+1$; go to Step 6.

Step 12. If the selected task is a Load (whose Make task does not require the robot) or an Unload, then:

Step 12.1. Process the task;

Step 12.2. Update the instant of time At_M at which the robot will be available again according to the processing time of the task executed; set t to this latter value;

Step 12.3. If a Load task has been selected, then update in FTA the earliest starting time of the Make task associated;

Step 12.4. If an Unload task has been selected, set to $t+1$ the time at which the machine which have processed its previous Make task will be available; update the instant of time At_M

at which the robot will be available again according to the processing time of the task executed; update t accordingly;

Step 12.5. Delete from LTA the processed task; go to Step 6.

Step 13. If the selected task is a Load task such that the following Make task requires the presence of the robot (as shown in list LTR), then:

Step 13.1. Process the task Load and, immediately after, the following Make task;

Step 13.2. Update the variable in list AT indicating when the robot will be available again (*i.e.*, after an interval of time equal to the sum of the processing times of the Load and the Make operations), while setting the availability of the machine which has processed the Make task equal to infinity after this Make operation has been performed;

Step 13.3. Update the variable in the FTA list of the earliest starting time of the corresponding Unload task, *i.e.*, t plus the processing times of the Load and the Make operations performed, say $p_{L_i^k}$ and $p_{M_i^k}$ respectively;

Step 13.4. Update $t = t + p_{L_i^k} + p_{M_i^k}$;

Step 13.5. Delete from LTA the Load task; delete from LT the Make operation.

Step 13.6. Go to Step 6.

Step 14. Return the makespan: $C_{\max} := t$

In Figure 4.4, we show an application of the $LMUA$ algorithm with 4 parts: Make operations for parts 1 and 3 require machine m_2; Make operations for part 2 and 4 require machine m_1; moreover, the Make operation for part 2, namely M_2^1, also requires resource R.

In order to determine the computational complexity of the $LMUA$ algorithm, note that, for a given instance of Z parts, there are $3Z$ tasks and:

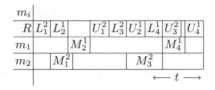

Fig. 4.4. An application of the $LMUA$ algorithm

Step 1: the lists can be constructed in $O(Z)$;
Step 2: the cycle **while** is repeated at most $3Z$ times, and the task selection requires at most $3Z$ comparisons;
Step 3: runs in $O(1)$;
Step 4: can be processed in $O(1)$;
Step 5: runs in $O(1)$;

Hence, for a given instance of the LMU problem, denoting the size of the input as Z, the $LMUA$ algorithm has a worst case complexity $O(Z^2)$.

4.5 Computational Results

In this section, we present some computational experiments with the $LMUA$ algorithm on randomly generated problems. We considered several cell configurations with $m \in [4, 10]$ machines and one robot tending all the machines. For each configuration, we considered an increasing number of jobs $n \in [10, 80]$. Note that, for instance, 80 jobs correspond to 240 tasks. Processing times for each loading and unloading operation are generated randomly, using a uniform distribution in the range of $[20 - 70]$ time units. Processing times of Make operations are generated randomly using a uniform distribution in the range of $[120 - 360]$ time units. We considered different scenarios associated with a probability p_r equal to 0, 0.1, 0.2, 0.3, and 0.4 that a generic part would require the robot during the Make operation. The algorithm implementation was done in a WINDOWS/C environment on an AMD Athlon PC running at 900 MHz.

Results are summarized in the following tables, which show, for each cell configuration, the values of the makespan depending on the number n of jobs. Each table is associated with a scenario corresponding to a probability p_r.

Let us first consider a scenario in which jobs do not require the resource robot during the Make operation (Table 4.1). Observe that, for a given m, the makespan increases as n increases, while it decreases, for a given n, as the number of machines m increases. To evaluate the trends in the makespan, we provide the chart in Figure 4.5.

Table 4.1. Scenario with $p_r = 0$

$p_r = 0$	←——Jobs (n) ——→							
Mac. (m)	**10**	**20**	**30**	**40**	**50**	**60**	**70**	**80**
4	1977.9	3265.5	4655.9	6774.6	8110.6	9300.4	10711.1	12077.3
6	1681.8	2733.4	4157.2	5460.4	6673.2	8117.8	9633.8	11220.8
8	1460	2712.1	3784.2	5400.4	6660	8014	9054.3	10482
10	1438.1	2678.2	3723.2	5314.5	6638.7	7754.5	8698.9	10458.3

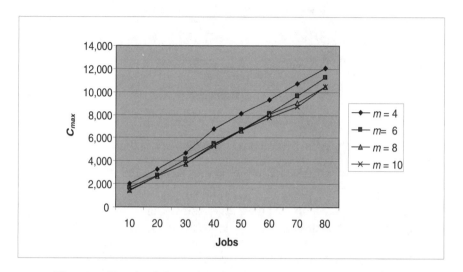

Fig. 4.5. Trends of the makespan as the number n of jobs increases

The makespan value, given a fixed number m of machines, seems to be linearly dependent on the number of jobs. For a certain range of the number of jobs ($n \in [10, 30]$), the trends are very similar, and it seems that the number m of machines does not affect C_{max}. Instead, as n increases the trends are well defined, and the difference between the makespan values is much more observable when m increases from 4 to 6 than when it increases from 6 to 8 or 10.

We now analyze the trends in the makespan with regard to the increase in the number m of the machines, for a given n. Figure 4.6 shows how C_{max} decreases proportionally as the number m increases. Moreover, from Table 4.1 it is easy to see that when the number of jobs increases from $n = 30$ to $n = 40$, the variation of the associated makespan values is higher than in the other cases.

Finally, we study what happens if the probability p_r that a job requires the robot during the Make operation is greater than zero. Tables 4.2, 4.3, 4.4 and 4.5 summarize such results.

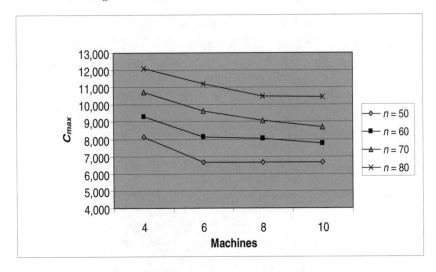

Fig. 4.6. Trends in the makespan as the number m of machines increases

Table 4.2. Scenario with $p_r = 0.1$

$p_r = 0.1$	←Jobs (n) →							
Mac. (m)	10	20	30	40	50	60	70	80
4	2069.9	3486.6	5118.2	7300	8923.4	9999.7	11690.6	13615.2
6	1747.8	3096.8	4536	6292.1	7830.5	9298.7	11094.8	13057.5
8	1534.3	3077.5	4487	6193.8	7510.2	8952.5	10382.5	12008.9
10	1506.4	2969.3	4322	6125.6	7503.8	8771.4	10138.7	11849.9

Table 4.3. Scenario with $p_r = 0.2$

$p_r = 0.2$	←Jobs (n) →							
Mac. (m)	10	20	30	40	50	60	70	80
4	2152.5	3674.4	5320.9	7808.6	9428.6	10962.4	12574.6	14435.3
6	1818.3	3458.4	5143.6	7109.3	8806.4	10647.6	12018.1	14319
8	1775.3	3373.7	4954	6761.6	8430.1	10336.6	11880	13752
10	1591.9	3367.3	4919.1	6651.7	8283.1	9914.6	11270.6	13356.7

Table 4.4. Scenario with $p_r = 0.3$

$p_r = 0.3$			←—**Jobs** (n) —→					
Mac. (m)	**10**	**20**	**30**	**40**	**50**	**60**	**70**	**80**
4	2301.7	3743.1	5611.7	8289	10223	11656.9	13882.7	15894.9
6	1929.2	3644.8	5462.2	7798.4	9358.6	11496.9	13212	15787.6
8	1821	3465.1	5440.8	7341.7	9207.9	10969.7	12786.3	15046.8
10	1760.1	3412.8	5092	7252.4	9048.8	10923.3	12089.6	14451

Table 4.5. Scenario with $p_r = 0.4$

$p_r = 0.4$			←—**Jobs** (n) —→					
Mac. (m)	**10**	**20**	**30**	**40**	**50**	**60**	**70**	**80**
4	2383	4134	6073.6	9009.7	10859.4	12837.6	14857.8	17176.5
6	2024.3	3960.9	5876.7	8243.3	10232.2	12491.2	14268.5	16624.1
8	1837.3	3671.6	5734.9	7969.9	9644.2	11775	13719.8	15950.3
10	1815.5	3603.2	5406.4	7700.9	9577.2	11424.9	13693.4	15296

Note that as p_r increases, C_{max} decreases proportionally. The chart in Figure 4.7 shows that the makespan values increase proportionally with probability p_r and the number of jobs n.

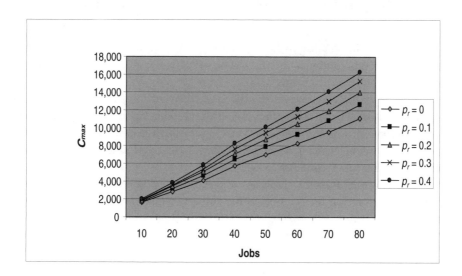

Fig. 4.7. Makespan values, for a given m, as n and p_r increase

The chart in Figure 4.8, instead, shows that the variation in the makespan, when probability p_r increases, is not proportional to the number m of machines, for a given n. In fact, it can be seen how the influence of p_r on the makespan tends to decrease as m increases.

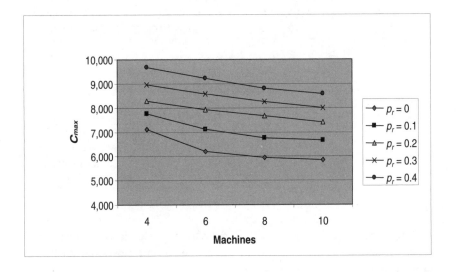

Fig. 4.8. Makespan values, for a given n, as m and p_r increase

Note that the maximum CPU time (in seconds) spent by the algorithm, obtained for the combination ($p_r = 0.4, m = 6, n = 80$), was 2.67, whereas the average running time was 0.38.

A final analysis is devoted to the possibility of improving the efficiency of the robotic cell. In particular, we examine whether it is more profitable to reduce the processing times of Make operations, improving machines efficiency, or to reduce the processing times for loading and unloading operations, modifying the robot configuration.

Firstly, we analyze for the case $p_r = 0$, what happens if the range of the uniform distribution for the processing times of Make operations is decreased by 30%, *i.e.*, from [120 − 360] to [120 − 268] time units. Table 4.6 summarizes the results obtained with this new scenario, and Table 4.7 shows the percentage of the reduction of C_{max}.

We now analyze what happens if the range of the uniform distribution for the processing times of Load and Unload operations is decreased by 30%, *i.e.*, from [20 − 70] to [20 − 55] time units. Table 4.8 summarizes the results obtained in this case, and Table 4.9 gives the percentage of the reduction of C_{max}.

Table 4.6. Reducing processing times of Make operations

$p_r = 0$				←—Jobs (n) —→				
Mac. (m)	**10**	**20**	**30**	**40**	**50**	**60**	**70**	**80**
4	1600.1	2781.3	3831.6	5735.3	6806.3	7877.1	8862.9	10284.7
6	1412.1	2481	3707.8	5022.4	6300.3	7630	8586.4	10150
8	1263.9	2409.8	3585.1	4948.5	6274.4	7480	8405.4	9343.2
10	1210.7	2392.1	3459.2	4918.7	6256.2	7290.3	8082.4	9900.3

Table 4.7. Percentage reduction of C_{max}

$p_r = 0$				←—Jobs (n) —→				
Mac. (m)	**10**	**20**	**30**	**40**	**50**	**60**	**70**	**80**
4	19.10%	14.83%	17.70%	15.34%	16.08%	15.30%	17.25%	14.84%
6	16.04%	9.23%	10.81%	8.02%	5.59%	6.01%	10.87%	9.54%
8	13.43%	11.15%	5.26%	8.37%	5.79%	6.66%	7.17%	10.86%
10	15.81%	10.68%	7.09%	7.45%	5.76%	5.99%	7.09%	5.34%

Table 4.8. Reducing processing times of Load and Unload operations

$p_r = 0$				←—Jobs (n) —→				
Mac. (m)	**10**	**20**	**30**	**40**	**50**	**60**	**70**	**80**
4	1833.9	3071.5	4312.1	6273.3	7639.3	8816.5	10015.8	11228.9
6	1567.7	2400.4	3607.3	4847.7	5467.2	6682	7717.9	9035.2
8	1334.5	2398.5	3264.2	4475.5	5458.9	6653.9	7527.2	8697.5
10	1301.5	2378.7	3138.8	4298.5	5429.5	6354	7097.6	8415.5

Table 4.9. Percentage reduction of C_{max}

$p_r = 0$				←—Jobs (n) —→				
Mac. (m)	**10**	**20**	**30**	**40**	**50**	**60**	**70**	**80**
4	7.28%	5.94%	7.38%	7.40%	5.81%	5.20%	6.49%	7.02%
6	6.78%	12.18%	13.23%	11.22%	18.07%	17.69%	19.89%	19.48%
8	8.60%	11.56%	13.74%	17.13%	18.03%	16.97%	16.87%	17.02%
10	9.50%	11.18%	15.70%	19.12%	18.21%	18.06%	18.41%	19.53%

It is easy to observe that a decrease in the processing times in both cases results in a reduction of C_{max} even if the latter is not proportional to the former. In fact, the maximum reduction obtained is 19.89% when the processing times of loading and unloading operations are decreased, and 19.10% when the processing times of Make operations are decreased instead.

4.6 Conclusions

We studied the general problem of sequencing multiple jobs where each job consists of multiple ordered tasks and task execution requires the simultaneous usage of several resources. The case of an automatic assembly cell is examined. The $\mathcal{NP}$-completeness in the strong sense of the problem is proved for an automatic assembly cell with four machines. A heuristic algorithm is proposed and we give computational results for an assembly cell considering different numbers of machines and one robot. The procedure at each iteration selects a task, based on the partial schedule obtained for the parts that have already been loaded by the assembly process. This feature of the proposed algorithm indicates that the approach presented can be applied in on-line scenarios as well as in a dynamic scheduling environment.

5

Tool Management on Flexible Machines

Flexible manufacturing systems (FMS) are production systems consisting of identical multipurpose numerically controlled machines (workstations), an automated material handling system, tools, load and unload stations, inspection stations, storage areas and a hierarchical control system. The main objective of these systems is to achieve the same efficiency as flow lines, while maintaining the same versatility as traditional job shops.

In this chapter, we focus on a cell manufacturing scenario characterized by a numeric control machine (NC, see Figure 5.1), capable of processing a set of jobs by means of a proper set of tools (see Figure 5.2). Since the latest generation of NC machines supports automatic tool changeovers and can be programmed in real time, a setup time must be considered between two changeovers. The latter features together with the automatic storage and loading of the parts, makes it possible to the parts to have a flexible flow. As a consequence, the family of parts can be produced simultaneously with positive effects on the inventory and on quick changes in the demand. The most important cases are those in which the setup times are not negligible with respect to the operation processing times, and thus, we focus our attention an the issue of minimizing the number of changeovers to reduce the overall production time.

5.1 Background

At the individual machine level tool management deals with the problem of allocating tools to a machine and simultaneously sequencing the parts to be processed so as to optimize some measure of production performance. This generic one-machine scheduling problem, or loading problem in the terminology of Stecke [163], can somehow be seen as the FMS analog of the fundamental one-machine scheduling problem of traditional manufacturing.

121

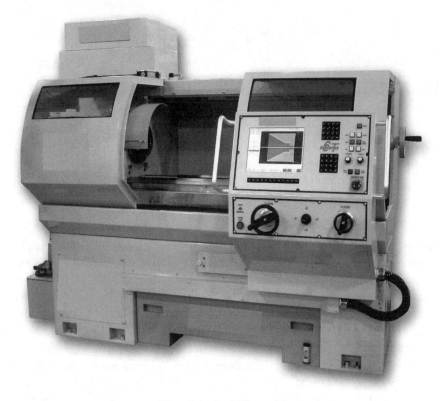

Fig. 5.1. An NC machine

A more precise formulation of the problem can be stated as follows. A part set or production order containing N parts must be processed, one part at a time, on a single flexible machine. Each part requires a subset of tools which have to be placed in the tool magazine of the machine before the part can be processed. The total number of tools needed to process all the parts is denoted as M. We represent these data as an $M * N$ tool-part matrix A, with $a_{ij} = 1$ if part j requires tool i, or 0 otherwise, for $i = 1, \ldots, M$ and $j = 1, \ldots, N$. The tool magazine of the machine features C tool slots. When loaded on the machine, tool i occupies s_i slots in the magazine ($i = 1, \ldots, M$). We assume that no part requires more than C tool slots for processing. We refer to the number C as the capacity of the magazine and to s_i as the size of tool i. (Typical magazine capacities lie between 30 and 120. Tool sizes are usually in the range of $\{1,2,3\}$, with tools of size 1 being most common.) The total number of tools required, *i.e.*, M, can be much larger than C, so that it is sometimes necessary to change tools while processing the order. A tool switch consists of removing one tool from the magazine and replacing it with

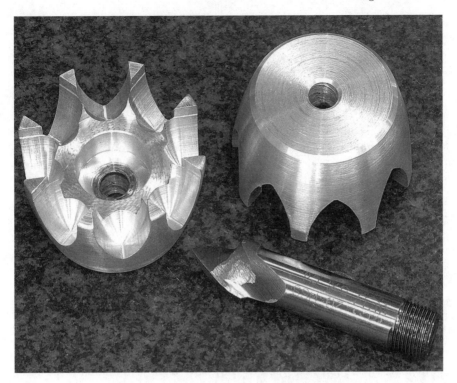

Fig. 5.2. Example of tools for NC machines

another one. A batch of parts is called feasible if it can be processed without any tool switches.

In some situations, the total number of tool switches incurred while processing an order appears to be a more relevant performance criterion than the number of switching instants (*i.e.*, the number of batches). This is for instance the case when the setup time of operations is proportional to the number of tool interchanges, or when the tool transportation system is congested.

In this section, we address the following tool switching problem: determine a part input sequence and an associated sequence of tool loadings such that all the tools required by the j-th part are present in the j-th tool loading and the total number of tool switches is minimized. In this form, the tool switching problem was investigated in many papers. All these papers are restricted to the special case where the tools occupy exactly one slot in the tool magazine. We shall assume that this condition holds throughout the section. (Note that the formulation of the problem becomes ambiguous when the assumption is lifted.) Crama *et al.* [58] proved that the tool switching problem is $\mathcal{NP}$hard for any fixed C. They also observed that deciding whether there exists a job sequence requiring exactly M tool setups is $\mathcal{NP}$-complete. The tool switching problem can be divided into two interdependent problems, namely:

1. part sequencing: determine an optimal part sequence, and
2. tooling: given a fixed part sequence, determine a tool loading sequence that minimizes the number of tool switches.

Tang and Denardo [164] established that the tooling subproblem can be solved in time $O(MN)$ by applying the so-called Keep Tool Needed Soonest (KTNS) policy. This policy establishes that, whenever tools must be removed from the magazine in order to make room for the tools required by the next part, the tools that are kept should be those that will be needed at the earliest time in the future.

The optimality of the KTNS principle was previously established by Belady [33] for a restricted version of the tooling problem. The Tang and Denardo's proof of correctness for the KTNS principle relies on ad hoc combinatorial arguments. Crama *et al.* [58] present a more compact proof based on an appropriate integer programming formulation of the tooling subproblem where the tooling subproblem is reducible to a network maximum flow problem, even in its generalized version where each tool i has its own setup time b_i and the objective is to minimize the sum of all setup times. When all setup times are equal, *i.e.*, when the objective is only to minimize the total number of switches, then the integer program can be solved by a greedy algorithm which turns out to be equivalent to the KTNS algorithm. The previous results have been further extended by Privault and Finke [146]. These authors give a direct network flow formulation of the tooling subproblem which allows them to model changeover costs in the form d_{ik} when loading tool i after unloading tool k. This approach leads to an $O(N^2)$ optimization algorithm for the generalized tooling subproblem. Similar reformulations have also been exploited by several authors in the context of lot-sizing with sequence-dependent changeover costs.

In spite of the simplicity of the tooling subproblem, the tool switching problem remains a hard one. Many heuristics have been proposed for its solution, but we are not aware of any successful attempts to reasonably solve large instances to optimality.

Heuristics, for the tool switching problem, come in two types: construction heuristics, which progressively construct a single, hopefully good part sequence, and local search heuristics, which iteratively modify an initial part sequence. In the first class, several approaches are based on approximate formulations of the tool switching problem as a travelling salesman problem, where the "distance" between two parts is an estimate of the number of tool switches required between these parts. It may be interesting to note that one of these travelling salesman formulations is in fact an exact model for a database management problem closely resembling the tool switching problem.

Another type of construction heuristics fall into the category of greedy heuristics: parts are successively added to a current subsequence on the basis of some (dynamically updated) priority criterion. Lofgren and McGinnis [123] develop a simple greedy heuristic that considers the "current" machine set-up, identifies the next job (printed circuit card - PCC) based on the similarity of its requirements to the current set-up, and then determines the components to be removed (if any) based on their frequency of use by the remaining PCCs. Rajkurmar and Narendran [148] present a greedy heuristic, which exploits the similarities between boards and current setup on the magazine by their similarity coefficients (see the next section for an example).

Various local search strategies (2-exchanges, tabu search, *etc.*) for the tool switching problem have been tested. Sadiq *et al.* [152] conduct a research on printed circuit boards (PCBs) sequencing and slot assignment problem on a single placement machine by the intelligent slot assignment (ISA) algorithm. Gronalt *et al.* [83] model the combined component loading and feeder assignment problem, called as the component-switching problem, as a mixed-integer linear program. A recursive heuristic is proposed to solve this combined problem. Maimon and Braha [126] develop a genetic algorithm and compare it with a TSP-based spanning tree algorithm. Recently, Djelab *et al.* [64] propose an iterative best insertion procedure using hypergraph representation to solve this scheduling problem. Some beginning efforts have been tried to handle the "non-uniform tool-sizes" issue using different formulations such as in [84, 131, 146]. Privault and Finke [146] extend the KTNS policy to take into account the "non-uniform tool sizes" by using the network flow formulation. Gunther *et al.* [84] extend the PCB assembly setup problem for single machine when the size of feeder required more slots in the magazine. They solve three sub-problems: board sequence, component loading and feeder assignment, sequentially by a TSP-based heuristic. Their heuristic first constructs a board sequence using an upper bound on component changeovers between two consecutive boards. The KTNS rule is implemented to evaluate the performance of the board sequence. Then an iterative procedure is developed using 2-opt heuristic to improve the previous solution. The drawback of this approach is that it does not consider the current status of the magazine. Recently, Matzliach and Tzur [131] have shown the complexity of this case to be $\mathcal{NP}$-complete. They also propose two constructive heuristics which provide solutions that are extremely closed to optimal solution (less than 2%). In the same streamline of this extension, Matzliach and Tzur [130] concern with another aspect of the tool switching problem when parts that need to be processed on the machine arrive randomly and tool sizes are non-uniform. In

another direction, Rupe and Kuo [151] relax the assumption of tool magazine capacity restriction. They allow job splitting and develop a so-called "get
tool needed soon" (GTNS) policy. The GTNS strategy is proved to give the
optimal solution for this case of the problem.

5.1.1 Definition of the Generic Instance

Let M be the number of jobs to be processed by an NC machine, which has
an automatic mechanism for tool change, and let N be the total number of
tools requested in order to process the entire set of the jobs. Assume that the
tool storage has $C < N$ slots. Moreover, the set of tools requested by job j
can be represented by means of a matrix $A = [a_{ij}]$ whose generic element a_{ij}
is:

$$a_{ij} = \begin{cases} 1 \text{ if tool } i \text{ is requested by job } j \\ 0 \text{ otherwise} \end{cases}$$

Given a sequence of jobs $\{j_1, j_2, \ldots, j_m\}$ and a set of N tools, the problem
is defined by matrix A, whose column vectors are a_j. Once A is defined [58], a
sequence of jobs can be seen as a permutation of the integer $1, 2, \ldots, m$ as well
as a permutation of the columns in A. Since the number of tools needed to
produce all the jobs is generally higher than the capacity of the tool storage,
it is sometimes necessary to change tools between two jobs in the sequence.

The tool switching problem can be redefined as follows: according to a certain sequence of jobs, find a matrix $P \in \{0,1\}^{N*M}$, obtained by permutating
the columns in A, and a matrix $T \in \{0,1\}^{N*M}$, containing C elements equal
to 1 in each column, such that $t_{kj}=1$ if $p_{kj}=1$, with the objective function of
minimizing the number of switches requested by the loading sequence:

$$T = \sum_{k=1}^{N} \sum_{j=2}^{M} (1 - t_{kj-1}) t_{kj}$$

5.1.2 Assumptions

The following hypothesis will be assumed in the rest of this chapter [28, 58,
164]:

1 The tool magazine is always full, loaded to the maximum of its capacity:
   the rationale behind this assumption is that having a lower load than the
   maximum allowed provides no advantages to the solution; moreover, the
   initial configuration (say, at the beginning of the working day) is assumed
   to be given at zero *setup* cost.

2   Each tool can be allocated to exactly one magazine slot. Removing this assumption may cause several problems (for example, deciding where to locate tools requiring more than a single slot).

3   The magazine accepts any configuration of tools: this allows us to always obtain the tool configuration most adequate to the specific problem instance.

4   The time needed to load a tool is constant, and equal to the unload time, and is not dependent on the specific job.

5   We can load only one tool at a time: this hypothesis is induced by the mechanism that delivers the tools from the magazine to the main machine.

6   The subset of tools required to process a job is known in advance (at the beginning of the working day).

7   The whole job list is known at the beginning of the working day.

Different formulations could also be obtained by changing the objective function. For instance, minimizing the number of tool switches is equivalent to minimizing the number of setups based on the equality reported below:

$$tool\ switches + c = setup\ number$$

where C denotes the magazine capacity.

5.2 The Binary Clustering and the KTNS Approaches

The Binary Clustering algorithm is based on a very simple idea: suppose job i requires a subset of the tools required by a certain job j. Putting j before i allows one not to change tools when going from the processing of j to the processing of i. This idea can be formalized as follows: we see the set of tools required for each job, and when there is a job whose set of tools is a subset of the set of tools required by another job, then we put it immediately after the latter job.

In the following, we provide a way of implementing this algorithm. Due to its nature this algorithm is also called Binary Clustering algorithm.

Assume that we have a tool-job matrix as reported in Table 5.1.

Table 5.1. A tool/job matrix

tool/job	1 2 3 4 5
1	1 0 1 0 0
2	1 1 1 0 0
3	0 0 0 0 1
4	0 1 0 1 0
5	1 0 1 1 0
6	0 0 1 0 1

Suppose that the tool machine is currently empty and that the capacity of the magazine is $C = 4$.

We order jobs according to the following criterion. We assign value 2^{M-k} to column k, and assign a value to each row by summing up all the values obtained in the columns if they correspond to an entry equal to 1. We then sort the rows according to non increasing values.

In Table 5.2 we give the score of each row. Table 5.3 shows the row ordered according to non increasing values.

Table 5.2. The score of each row

tool/job	1 2 3 4 5	Score
1	1 0 1 0 0	20
2	1 1 1 0 0	28
3	0 0 0 0 1	1
4	0 1 0 1 0	10
5	1 0 1 1 0	22
6	0 0 1 0 1	5

We do the same for the columns, assigning to each row k a value 2^{N-k}. In Table 5.4 we show the score of each column, while Table 5.5 gives the column ordered according to non increasing values.

We now iterate the ordering process until no changes is possible in both rows and columns. The final matrix is given in Table 5.6.

Thus, according to this heuristic we can conclude that the job order is

$$3, 1, 2, 5, 4.$$

Table 5.3. The matrix ordered with respect to the rows

tool/job	1 2 3 4 5
2	1 1 1 0 0
5	1 0 1 1 0
1	1 0 1 0 0
4	0 1 0 1 0
6	0 0 1 0 1
3	0 0 0 0 1

Table 5.4. The scores of each column

tool/job	1	2	3	4	5
2	1	1	1	0	0
5	1	0	1	1	0
1	1	0	1	0	0
4	0	1	0	1	0
6	0	0	1	0	1
3	0	0	0	0	1
Score	56	36	58	20	3

Table 5.5. The matrix ordered with respect to the columns

tool/job	3 2 1 4 5
2	1 1 1 0 0
5	1 1 0 1 0
1	1 1 0 0 0
4	0 0 1 1 0
6	1 0 0 0 1
3	0 0 0 0 1

With this order, we first load tools 2, 5, 1, and 6, then, when job 2 requires tool 4, applying the KTNS rule, we can unload tool 1 which is no longer requested by the successive jobs, and, when job 4 requires tool 3, we remove one among tools 2, 4, and 5.

Table 5.6. The final matrix ordered

tool/job	3 1 2 5 4
2	1 1 1 0 0
5	1 1 0 1 0
1	1 1 0 0 0
6	1 0 0 0 1
4	0 0 1 1 0
3	0 0 0 0 1

5.3 The Proposed Algorithms

5.3.1 Algorithm 1

The first algorithm we propose creates first groups of compatible jobs and then utilizes a sequencing heuristics. The creation of compatible groups is done by finding maximum cardinality matchings on graphs in an iterative fashion. The algorithm is sketched in the following.

Step 0. Initialize a vector fix to $\emptyset$.

Step 1. Build a graph $G(V, E)$, with vertex set V equal to the job set and edges set E such that an edge $(i, j) \in E$, if the cardinality of the set of tools required simultaneously by i and j is less than C.

Step 2. Find a maximum cardinality matching in G. If $\mathcal{M} = \emptyset$ (*i.e.* $E = \emptyset$) go to Step 4, otherwise go to Step 3.

Step 3. Build a new graph $G'(V', E')$ starting from $G(V, E)$ in the following way. Each node i of V' corresponds to either:
1. one edge of $\mathcal{M}$. In this case i can be seen as a new job requiring a set of tools given by the union of the tools requested by the two jobs associated with the edge of $\mathcal{M}$, or
2. one node of V, not covered by the matching. In this case, the job requires the same tools as those required in V.

Two nodes are made adjacent in V' following the same relation used in V. Rename $G'(V', E')$ as $G(V, E)$ and go back to Step 2.

Step 4. Increase the magazine capacity C by one. If the capacity is less than or equal to N, go to Step 5, otherwise go to Step 6.

Step 5. Determine the new graph $G(V, E)$ according to the new capacity, and add all the edges in fix. Go to step 4.

Step 6. From fix, generate two other vectors, denoted as fix' and fix'', by respectively reversing fix, and randomly reordering the sublists in fix associated with edges belonging to the same subgraph.

Step 7. For each vector fix, fix' and fix'', determine some paths having as their first node the node representative of group j (with $j = 1, 2, \ldots, M$), respectively. The strategy to obtain the paths is described in the following sequencing algorithm.

The sequencing subroutine

Step 7.1. Examine the vector of edges and find the starting edge, that is the one having one node equal to node j. Label this edge as visited; add this edge to the list of path vectors, by labelling the two nodes as extremes.

Step 7.2. Start from the first element of the vector and examine these elements that have not yet been visited one by one. As soon as there is an edge having one node equal to one of the extremes which does not generate cycles, insert this edge into the list of the path vector.

Step 7.3. If there are nodes that have not yet been visited, go to Step 7.2, otherwise go to Step 7.4.

Step 7.4. Examine the path vector, and, by ordering its elements, determine an acyclic path on the graph.

End of the sequencing subroutine

Step 8. Examine all the paths obtained in Step 7 and, once a saturation of the magazine capacity is obtained for each path, determine for each path the tool switches needed to obtain such a sequencing. In the following, we will denote a sequence as optimal if it minimizes the number of changes.

The reason for adopting a set covering approach to determine the groups is that it allows us to obtain a group of mutually compatible groups rather than having only couples of compatible elements.

In order to determine the number of tool switches, it has been assumed that from the very beginning of the process, the tools required by the first job have already been installed, and, if there is a residual capacity of the magazine, it is also assumed that the remaining slots have been loaded with the tools required by the successive jobs, until it is full.

Example of an Application of Algorithm 1

Consider the tool/job matrix represented in Table 5.7.

Table 5.7. Tool/job matrix

tool/job	1	2	3	4	5	6	7	8	9	10
1	0	1	0	0	0	0	0	0	0	0
2	1	0	0	0	1	1	1	0	0	0
3	0	1	0	0	1	1	0	0	0	0
4	0	0	1	1	0	0	1	0	1	0
5	0	0	0	0	1	1	0	0	0	0
6	1	0	0	0	0	0	0	1	0	0
7	0	0	1	0	1	0	0	0	0	1
8	0	0	0	1	0	0	0	1	0	1
9	0	1	0	0	0	1	1	1	1	0
10	0	0	0	0	0	0	0	0	1	0

In Step 1, the compatibility graph $G(V, E)$ is built (see Figure 5.3); while in Step 2 we compute a matching given in Figure 5.4.

Grouping the matching nodes (Step 3) allows us to obtain a graph without edges (see Figure 5.5, where the labels in brackets refer to the nodes of the graph of the previous step). In Figure 5.6, we depicted the iterated execution of Steps 4 and 5, by using different kinds of edge representations to cope with the different values of the capacity C.

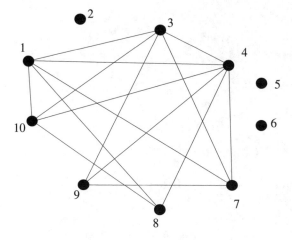

Fig. 5.3. The compatibility graph

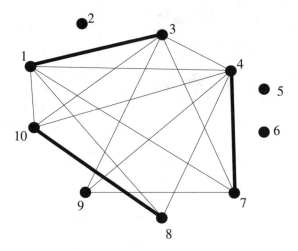

Fig. 5.4. A matching in the compatibility graph

Starting from the graph built in Step 5, we have the following vector fix containing the edges of the graphs as shown in Figure 5.6:

$$[(2,5),(2,7),(3,7),(4,5),(1,3),(1,4),(1,6),(1,7),$$
$$(2,3),(2,4),(2,6),(3,5),(3,6),(5,7),(6,7),(1,2),$$

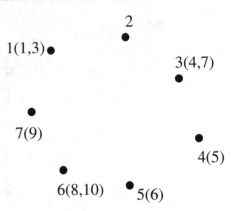

Fig. 5.5. The new graph G'

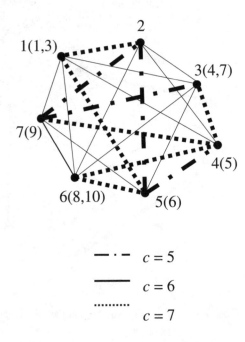

Fig. 5.6. Iterated Step 4 and Step 5

$$(1,5),(3,4),(4,6),(4,7),(5,6)]$$

with a possible reordering given by the following vector:

$$[(4,5),(3,7),(2,7),(2,5),(6,7),(5,7),(3,6),(3,5),$$
$$(2,6),(2,4),(2,3),(1,7),(1,6),(1,4),(1,3),(5,6),$$
$$(4,7),(4,6),(3,4),(1,5),(1,2)].$$

An example of the path vector in Step 7, obtained by considering only the first of the above reported vectors is:

$$[(1,3),(3,7),(2,7),(2,5),(4,5),(4,6)]$$

and from this vector we obtain the path:

$$1,3,7,2,5,4,6$$

whose cost is 11.

By iterating the above step (see again the algorithm), one can find the sequence that minimizes the number of tool switches, whose value is 9.

5.3.2 Algorithm 2

Algorithm 2 uses a heuristic to build groups that consists of determining, among the set of tools, those that are most frequently requested. The heuristic proceeds by making successive reductions of the initial tool/job matrix trying to identify those tools that have the largest probability of being requested by successive jobs. The full description is described in the following.

Step 0. Let C' be the residual capacity of the magazine. Set $C' = C$. Consider the current tool/job matrix, and check whether the number of tools required by all the jobs is greater than the magazine capacity C. If this is the case, then go to Step 1; otherwise, go to Step 2.

Step 1. Find the tool which is required by the largest number of jobs (in case of a tie choose the tool arbitrarily). Restrict the tool/job matrix to those jobs requiring the latter tool, and set to zero all the entries of the row corresponding to this tool. Decrease by one the residual capacity C'; if $C' > 0$, go to Step 0, otherwise, go to Step 2.

Step 2. The configuration of tools to be installed is given by the set of tools selected in Step 1, plus those of the current jobs-tools matrix.

Step 3. On the original tool-job matrix assign, 0 to all entries of each column corresponding to a job requiring a superset of the set of tools found in Step 1. If there exists at least one entry different from zero, go to Step 0; otherwise, go to Step 4.

Step 4. Examine the sequence of tool configurations obtained with the iterative restrictions. Build a new, complete, weighted graph as follows. The node set is the set of tool configurations obtained by the algorithm. The length (weight) of each edge is determined by the cardinality of the intersection of the configurations associated with the edge. If one of the two configurations has a cardinality smaller than the magazine capacity, then the weight is increased by the difference between the magazine capacity and this cardinality. If both the configurations are smaller than the magazine capacity, then the weight is increased by the difference between the magazine capacity and the minimum configuration cardinality.

Step 5. Order the edges in a vector in non increasing values of the length. Initialize a counter i to 1.

Step 6. Examine the edge corresponding to i; label it as visited and insert it into a vector v; mark each node of the edge.

Step 7. Scan all the elements in the edge vector until the first edge which has one of its nodes in common with the edges in vector v and does not generate cycles.

Step 8. Add the edge so found in the path vector, and mark the node that has not yet been marked. If there are nodes that have not yet been visited go to Step 7, otherwise, go to Step 9.

Step 9. Compute a path exploiting the edges in the path vector.

Step 10. Compute the number of tools required to sequence the configurations, making sure that there are always C tools in the magazine.

Step 11. Update the counter i. If i is smaller than the cardinality of the elements contained in the edge vector, delete all entries in the path vector and go to Step 6; otherwise, go to Step 12.

Step 12. Among all the sequences so obtained, choose the one with the minimum number of tool changes.

Example of an Application of Algorithm 2

Let us consider the instance in the previous example. Given the tool/job matrix, the overall number of requested tools is greater than the storage capacity $C = 4$. Thus, we apply Step 1 of the algorithm, which returns tool 9 as the most used, as depicted in Table 5.8.

Table 5.8. Step 1: first iteration

tool/job	1	2	3	4	5	6	7	8	9	10	tot
1	0	1	0	0	0	0	0	0	0	0	1
2	1	0	0	0	1	1	1	0	0	0	4
3	0	1	0	0	1	1	0	0	0	0	3
4	0	0	1	1	0	0	1	0	1	0	4
5	0	0	0	0	1	1	0	0	0	0	2
6	1	0	0	0	0	0	0	1	0	0	2
7	0	0	1	0	1	0	0	0	0	1	3
8	0	0	0	1	0	0	0	1	0	1	3
9	0	1	0	0	0	1	1	1	1	0	5
10	0	0	0	0	0	0	0	0	1	0	1

At this point, the algorithm keeps on iterating between Step 0 and Step 1 as illustrated in Tables 5.9-5.11, producing tool 2, 3 and 5, respectively, as the most used. At this point, the residual capacity $C' = 0$, and Step 2 is invoked.

The configuration of tools to be installed is 2, 3, 5, and 9. With this configuration, only job 6 can be processed and the new tool/job matrix is the one represented in Table 5.12.

Table 5.9. Step 1: second iteration

tool/job	2	6	7	8	9	tot
1	1	0	0	0	0	1
2	0	1	1	0	0	2
3	1	1	0	0	0	2
4	0	0	1	0	1	2
5	0	1	0	0	0	1
6	0	0	0	1	0	1
7	0	0	0	0	0	0
8	0	0	0	1	0	1
9	0	0	0	0	0	0
10	0	0	0	0	1	1

Table 5.10. Step 1: third iteration

tool/job	6	7	tot
1	0	0	0
2	0	0	0
3	1	0	1
4	0	1	1
5	1	0	1
6	0	0	0
7	0	0	0
8	0	0	0
9	0	0	0
10	0	0	0

In Step 3, the algorithm proceeds with successive iterations which lead to the following tool configurations:

$$\{(2,3,5,9),(2,4,9,10),(2,3,5,7),(4,8),(2,6,8,9),(4,7,8),(1,3,9)\}$$

In Step 4, calculating the intersections among configurations, we proceed by determining the matrix of the intersection (see Table 5.13) and the corresponding graph (see Figure 5.7).

When Step 5 is executed, the adjacency vector is:

Table 5.11. Step 1: fourth iteration

tool/job	6	tot
1	0	0
2	0	0
3	0	0
4	0	0
5	1	1
6	0	0
7	0	0
8	0	0
9	0	0
10	0	0

Table 5.12. Tool/job matrix after Step 2

tool/job	1	2	3	4	5	6	7	8	9	10
1	0	1	0	0	0	0	0	0	0	0
2	1	0	0	0	1	0	1	0	0	0
3	0	1	0	0	1	0	0	0	0	0
4	0	0	1	1	0	0	1	0	1	0
5	0	0	0	0	1	0	0	0	0	0
6	1	0	0	0	0	0	0	1	0	0
7	0	0	1	0	1	0	0	0	0	1
8	0	0	0	1	0	0	0	1	0	1
9	0	1	0	0	0	0	1	1	1	0
10	0	0	0	0	0	0	0	0	1	0

$$[(4,6),(1,3),(1,7),(2,4),(4,5),(1,2),(1,4),(1,5),$$
$$(2,5),(2,6),(2,7),(3,4),(3,6),(3,7),(4,7),(5,6),$$
$$(5,7),(1,6),(2,3),(3,5),(6,7)].$$

The first couple selected by Step 6 is $(4,6)$; and at Steps 7 and 8, we obtain by successive scans of the edge vector, the path vector:

$$[(4,6),(2,4),(1,2),(1,3),(3,7),(5,6)].$$

Step 9 obtains form the path vector the sequence:

Table 5.13. Intersections among configurations

configurations	1	2	3	4	5	6	7
1	0	2	3	2	2	1	3
2	2	0	1	3	2	2	2
3	3	1	0	2	1	2	2
4	2	3	2	0	3	4	2
5	2	2	1	3	0	2	2
6	1	2	2	4	2	0	1
7	3	2	2	2	2	1	0

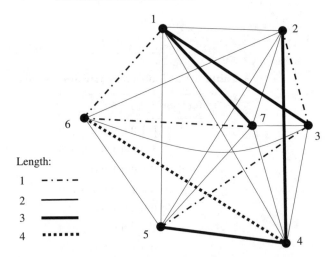

Fig. 5.7. Graph of the intersections among configurations

$$5, 6, 4, 2, 1, 3, 7,$$

Step 10 computes the cost of the configuration as equal to 9 and Steps 11 and 12 determine the best solution:

$$5, 4, 2, 7, 1, 3, 6,$$

whose cost is 8.

5.3.3 Algorithm 3

This heuristic also is based on first grouping and then searching for a specific path in an ad-hoc graph. The algorithm is described below.

Step 0. Let C' be the residual capacity of the magazine. Set $C' = C$. Consider the current tool/job matrix, and check whether the number of tools required by all the jobs is greater than the magazine capacity C. If this is the case, then go to Step 1; otherwise go to Step 2.

Step 1. Find the tool which is required by the largest number of jobs (in case of a tie, choose the tool arbitrarily). Restrict the tool/job matrix to those jobs requiring the latter tool, and set to zero all the entries of the row corresponding to this tool. Decrease by one the residual capacity C'; if $C' > 0$, go to Step 0, otherwise, go to Step 2.

Step 2. The configuration of tools to be installed is given by the set of tools selected in Step 1 plus those of the current tool/job matrix.

Step 3. In the original tool/job matrix, assign 0 to all the entries of each column corresponding to a job requiring a superset of the set of tools found in Step 1. If there exists at least one entry different from zero, go to Step 0; otherwise, go to Step 4.

Step 4. Examine the sequence of tool configurations obtained by the iterative restrictions. Build a new complete weighted graph as follows. The node set is the set of tool configurations obtained by the algorithm. The length (weight) of each edge is determined by the cardinality of the intersection of the configurations associated with the edge. If one of the two configurations has a cardinality smaller than the magazine capacity, then the weight is increased by the difference between the magazine capacity and this cardinality. If both the configurations are smaller than the magazine capacity then the weight is increased by the difference between the magazine capacity and the minimum configuration cardinality. Initialize counter i to 0.

Step 5. Consider the node of the graph corresponding to the counter i; insert it into the path vector; among all its successors, look for the one connected to the edge of the largest length.

Step 6. Consider the node indicated by the counter i, disconnect the node previously considered and if there is at least one edge in the graph, go to Step 5, otherwise, go to Step 7.

Step 7. The node sequence in the vector defines a path. Compute the sequencing needed to always have a magazine capacity of C tools. Increment the counter i by one, and if it is less than the bound, go to Step 5, otherwise, go to Step 8.

Step 8. Restore the original matrix of distances.

Step 9. Visit all the nodes according to their increasing numbering. Insert each node into the center of the path formed by the two edges obtained, considering the two successors with the heaviest weight. Remove the columns corresponding to the visited nodes from the matrix. Set the counter to 1.

Step 10. For each external node of the i-th triple, visit the successors of the node choosing the edges with the heaviest weight and insert them as external nodes into the current sequence. Remove all the edges adjacent to the currently visited nodes.

Step 11. If there are nodes that have not yet been visited, go to Step 10; otherwise, go to Step 12.

Step 12. Compute the number of tools needed to go from one configuration to the successive one, making sure there are always C tools in the magazine. Update the counter i and if it is less than the bound, go to Step 10, otherwise, go to Step 13.

Step 13. Among all the sequencings obtained with the two techniques, choose the one requiring the smallest number of tool changes.

The algorithm for forming the groups is the one used by Algorithm 2.

For the sequencing, two different approaches based on the highest number of intersections have been proposed. The first one, starting from the job at the *beginning* of the sequence, adds one job at a time trying to obtain the highest number of tools in common with the current job.

The second one, starts with a job at the *center* of the sequence, and adds nodes to the left and to the right of the current one, always choosing the one with the maximum number of common tools criterion. By iterating this step, new jobs are added to both sides of the sequence if the two successors have different weights, or only to the right side, if both weights are equal. After the computation of the number of tool changes, we choose the solution associated with the best strategy.

Example of an Application of Algorithm 3

We consider the same instance as the one in Example 1. As previously stated, the present algorithm utilizes the same sequencing technique as the one in Algorithm 2, so we will skip the Steps from 0 to 4. In particular, the starting point will be the intersection matrix and the associated graph (see Figure 5.7). At Step 5, node 1 is examined and node 3 is chosen among its successors.

Step 6: see Table 5.14 and Figure 5.8.

Table 5.14. Intersections among configurations

configurations	1	2	3	4	5	6	7
1	0	0	0	0	0	0	0
2	0	0	1	3	2	2	2
3	0	1	0	2	1	2	2
4	0	3	2	0	3	4	2
5	0	2	1	3	0	2	2
6	0	2	2	4	2	0	1
7	0	2	2	2	2	1	0

Step 5: the sequence becomes 1 3. Among the successors of 3, we choose 4.

Step 6: see Table 5.15 and Figure 5.9.

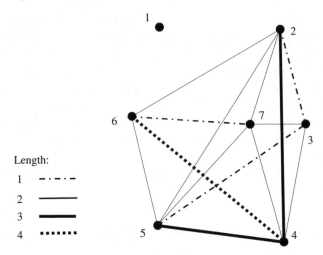

Fig. 5.8. Graph of the intersections among configurations

Table 5.15. Intersections among configurations

configurations	1	2	3	4	5	6	7
1	0	0	0	0	0	0	0
2	0	0	0	0	2	2	2
3	0	0	0	0	0	0	0
4	0	0	0	0	0	0	0
5	0	2	0	0	0	2	2
6	0	2	0	0	2	0	1
7	0	2	0	0	2	1	0

Step 5: the sequence becomes 1 3 4. Among the successors of node 3, node 6 is chosen.

Step 6: see Table 5.16 and Figure 5.10.

Step 5: the sequence becomes 1 3 4 6. Among the successors of 5, node 5 is chosen.

Step 6: see Table 5.17 and Figure 5.11.

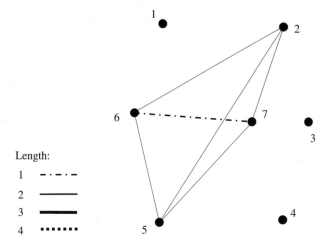

Fig. 5.9. Graph of the intersections among configurations

Table 5.16. Intersections among configurations

configurations	1 2 3 4 5 6 7
1	0 0 0 0 0 0 0
2	0 0 0 0 2 2 2
3	0 0 0 0 0 0 0
4	0 0 0 0 0 0 0
5	0 2 0 0 0 2 2
6	0 2 0 0 2 0 1
7	0 2 0 0 2 1 0

Step 5: the sequence becomes 1 3 4 6 2 5. Successor 7 is chosen.

Step 6: see Table 5.18 and Figure 5.12.

Step 7: now the sequence is:

$$1\ 3\ 4\ 6\ 2\ 5\ 7.$$

Successively, similar iterations are performed and the tool changes are computed from the remaining nodes of the graph in non decreasing order according to their numbering. Then, we proceed to Step 9.

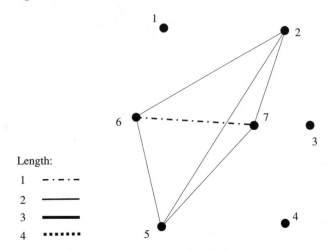

Fig. 5.10. Graph of the intersections among configurations

Table 5.17. Intersections among configurations

configurations	1 2 3 4 5 6 7
1	0 0 0 0 0 0 0
2	0 0 0 0 2 0 2
3	0 0 0 0 0 0 0
4	0 0 0 0 0 0 0
5	0 2 0 0 0 0 2
6	0 0 0 0 0 0 0
7	0 2 0 0 2 0 0

Step 9: starting from the situation in Figure 5.13, all nodes are re-examined in non decreasing order and for each one of them, the two biggest successors are considered; then, they are added to the right and the left of the node, respectively, obtaining the triples:

$$(3\ 1\ 7), (4\ 2\ 1), (1\ 3\ 4), (6\ 4\ 2), (4\ 5\ 1), (4\ 6\ 2), (1\ 7\ 2).$$

Step 10: starting from the triple associated with node 1, one proceeds according to the graph in Figure 5.14 and Table 5.19, associating to the more external nodes (identified by the squared frame in the figure), the nodes connected with the heaviest weight, (in our case, nodes 4 and 2) obtaining the sequence:

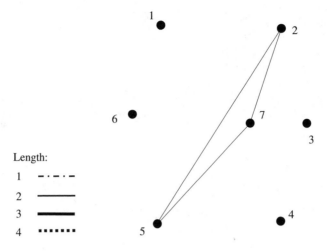

Fig. 5.11. Graph of the intersections among configurations

Table 5.18. Intersections among configurations

configurations	1	2	3	4	5	6	7
1	0	0	0	0	0	0	0
2	0	0	0	0	2	0	2
3	0	0	0	0	0	0	0
4	0	0	0	0	0	0	0
5	0	2	0	0	0	0	2
6	0	0	0	0	0	0	0
7	0	2	0	0	2	0	0

4 3 1 7 2.

Step 11: there are still nodes to be visited, go to Step 10.

Step 10: see Figure 5.15. Executing the step, we obtain

6 4 3 1 7 2 5.

Step 12: the sequence cost is 8. The successive increments of counter i give further iterations not reported for the sake of brevity.

Step 13: after the successive iteration has been computed, we choose the best result of the algorithm, which is 8.

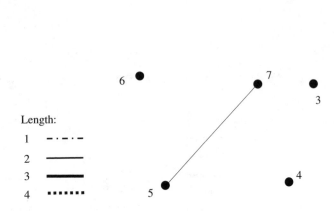

Fig. 5.12. Graph of the intersections among configurations

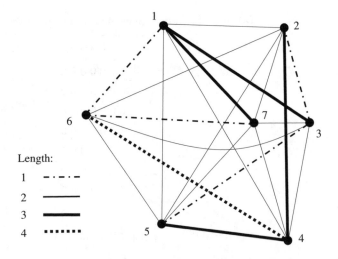

Fig. 5.13. Graph of the intersections among configurations

5.3.4 Algorithm 4

The algorithm is described in the following steps.

> **Step 1.** From the tool/job matrix, build the graph $H(V, E)$ with V corresponding to the tool set and, with an edge between two nodes if the two tools have at least one job in common.

Table 5.19. Intersections among configurations

configurations	1	2	3	4	5	6	7
1	0	2	3	2	2	1	3
2	2	0	1	3	2	2	2
3	3	1	0	2	1	2	2
4	2	3	2	0	3	4	2
5	2	2	1	3	0	2	2
6	1	2	2	4	2	0	1
7	3	2	2	2	2	1	0

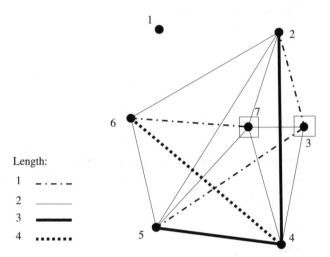

Fig. 5.14. Step 4: graph of the intersections among configurations

Step 2. Compute the out-degree (number of incident edges) for each node of V.

Step 3. Find the node with the minimum out-degree, remove the edges and store them in a matrix. If there are no isolated nodes, go to Step 2, otherwise, go to Step 4.

Step 4. Examine the matrix where the edges removed in the previous step have been stored, and for each column, list the jobs that can be processed adopting that associated configuration.

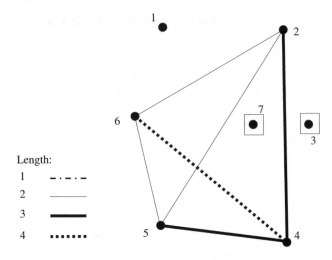

Fig. 5.15. Graph of the intersections among configurations

Step 5. Change the ordering of the columns of the matrix, so that the first successor of a job has the largest number of tools in common with the previous job. Ensure that the number of tools in the magazine is always kept equal to C, looking at the successive tools requested, and maintaining those already loaded following the KTNS strategy.

Step 6. Compute the removed tools.

As opposed to the algorithm presented described before, this algorithm does not require a grouping phase. The reason is because its strategy is based on the selection of the less requested tools, implemented by means of graph H, representing tools instead of jobs such as in graph G adopted by the previous algorithms.

Example of an Application of Algorithm 4

Adopting the same instance as the one in Example 1, we have:

Step 1: the graph $H(V, E)$ is given in Figure 5.16.

Step 2: the out-degree values are as follows:

$$2\ 6\ 5\ 5\ 4\ 3\ 5\ 4\ 8\ 2.$$

Step 3: select tool 1 and the edges 3 e 9.

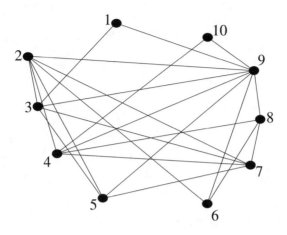

Fig. 5.16. Graph $H(V, E)$

Step 2: the graph is the one represented in Figure 5.17, whose out-degrees have cardinalities

0 6 4 5 4 3 5 4 7 2.

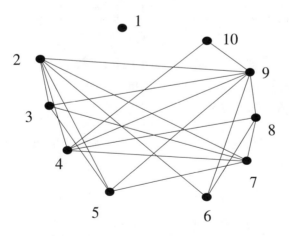

Fig. 5.17. First iteration of Step 2

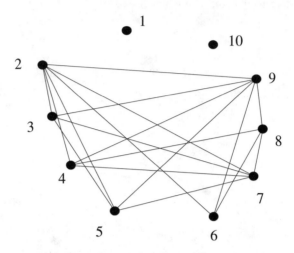

Fig. 5.18. Second iteration of Step 2

Going back to Step 2, node 10 is chosen and its adjacent nodes are 4 and 9. Thus, the algorithm proceeds in successive iterations until the matrix of removed edges is obtained (left side of Table 5.20).

Step 3: see Table 5.20.

Table 5.20. Complete table of removed edges according to the out-degree values

$$
\begin{array}{l}
1\ 0\ 0\ 0\ 0\ 0\ 0\ 0\ 0\ 1\,|\,1\ 0\ 0\ 0\ 0\ 0\ 0\ 0\ 0 \\
0\ 0\ 1\ 0\ 1\ 1\ 1\ 0\ 0\ 0\,|\,0\ 0\ 1\ 0\ 0\ 0\ 0\ 1\ 1\ 1 \\
1\ 0\ 0\ 0\ 0\ 1\ 1\ 1\ 0\ 0\,|\,1\ 0\ 0\ 0\ 0\ 0\ 0\ 0\ 1\ 1 \\
0\ 1\ 0\ 1\ 1\ 0\ 0\ 0\ 0\ 0\,|\,0\ 1\ 0\ 0\ 1\ 1\ 0\ 1\ 0\ 0 \\
0\ 0\ 0\ 0\ 0\ 1\ 1\ 1\ 1\ 0\,|\,0\ 0\ 0\ 0\ 0\ 0\ 0\ 0\ 1\ 1 \\
0\ 0\ 1\ 0\ 0\ 0\ 0\ 0\ 0\ 0\,|\,0\ 0\ 1\ 1\ 0\ 0\ 0\ 0\ 0\ 0 \\
0\ 0\ 0\ 1\ 1\ 1\ 0\ 0\ 0\ 0\,|\,0\ 0\ 0\ 0\ 1\ 0\ 1\ 0\ 1\ 0 \\
0\ 0\ 1\ 1\ 0\ 0\ 0\ 0\ 0\ 0\,|\,0\ 0\ 0\ 1\ 0\ 1\ 1\ 0\ 0\ 0 \\
1\ 1\ 1\ 1\ 1\ 0\ 1\ 1\ 1\ 0\,|\,1\ 1\ 0\ 1\ 0\ 0\ 0\ 1\ 0\ 1 \\
0\ 1\ 0\ 0\ 0\ 0\ 0\ 0\ 0\ 0\,|\,0\ 1\ 0\ 0\ 0\ 0\ 0\ 0\ 0\ 0 \\
\end{array}
$$

Step 4: the successive visit of the tool-job matrix leads to the matrix at the right side of Table 5.20, representing the sequencing.

Step 5: the improvement step returns the matrix in Table 5.21.

Table 5.21. Matrix obtained in Step 5

1 0 0 0 0 0 0 0 0 0
0 1 1 1 0 0 1 0 0 0
1 1 1 0 0 0 0 0 0 0
0 0 0 0 0 1 1 1 1 0
0 1 1 0 0 0 0 0 0 0
0 0 0 1 1 0 0 0 0 0
0 0 1 0 0 0 0 1 0 1
0 0 0 0 1 0 0 0 1 1
1 1 0 0 0 1 1 1 0 0
0 0 0 0 0 1 0 0 0 0

Step 6: ensuring a constant level of tools in the magazine, we calculate that the number of tool changeovers is 9.

5.4 Computational Analysis

To the best of our knowledge, no exact approach for this problem exists in the literature. Moreover, most of the contributions do not provide comparative analysis on the performance of the different approaches. The proposed algorithms have been implemented in C++ and tested on several instances (see the following sections for details on the test instances used).

Our main objective is the analysis of solution quality applying the same approach used in [58, 164].

5.4.1 Comparison with Tang and Denardo

In [164], the matrices are generated in the following way:

1. There are four cases. The number of jobs to be processed on an NC machine is $M = 10, 20, 30, 40$ and the number of required tools is $N = 10, 15, 25, 30$. For each one of these configurations the magazine capacity is $C = 4, 8, 10, 15$.

2. For each one of the first three cases, there are 5 tool/job randomly generated A_{ij} matrices. The number of tools per job is randomly generated with a uniform distribution in the interval $[1, C]$. For the fourth case, two matrices are generated instead of five.

Tests in [164] are performed applying KTNS strategy, in the first two cases on sequences of 100 jobs, and in the second two cases, on sequences of 200 jobs. The solutions found are given in Table 5.22.

Table 5.22. Performance of the four algorithms proposed

	Algo 1		Algo 2		Algo 3		Algo 4	
(N, M, C)	10/10/4	25/30/10	10/10/4	25/30/10	10/10/4	25/30/10	10/10/4	25/30/10
Instance 1	8	71	10	59	10	63	10	63
2	8	64	8	65	6	65	11	60
3	10	64	8	48	6	70	9	64
4	9	66	9	64	9	61	8	62
5	10	61	8	59	10	64	11	62
Average	9.0	65.2	8.6	59.0	8.2	64.6	9.8	62.2
(N, M, C)	15/20/8	30/40/15	15/20/8	30/40/15	15/20/8	30/40/15	15/20/8	30/40/15
Instance 1	24	102	24	108	22	83	30	94
2	27	107	16	105	29	94	20	101
3	27	-	27	-	21	-	30	-
4	23	-	25	-	20	-	25	-
5	24	-	25	-	20	-	24	-
Average	25.0	104.5	23.4	106.5	22.4	88.5	25.8	97.5

The average values of the above results are compared with those presented in [164] obtaining the chart in Figure 5.19.

From the chart it appears that even though the proposed algorithms require fewer iterations, they provide better solutions than those in [164]. The algorithm that seems to perform best is Algorithm 3, especially for larger values of the number of tools.

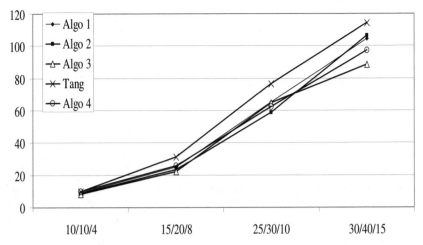

Fig. 5.19. Comparison with Tang and Denardo [164]

5.4.2 Comparison with Crama *et al.*

Test Problems

In this section, each heuristic algorithm is tested on 160 instances of the tool-switching problem, as done in [58]. The tool/job matrices are generated as described below.

Each randomly generated instance falls to one out of 16 types of instances characterized by a matrix dimension (N, M). Conventionally, each instance is denoted as a triple (N, M, C). There are 10 instances for each type. The $N * M$ matrices have values $(10, 10), (20, 15), (40, 30), (60, 40)$.

For each matrix dimension, an interval of parameters (min, max) with the following characteristics is defined:

$$min = minimum \ number \ of \ tools \ required \ to \ process \ a \ job.$$

$$max = maximum \ number \ of \ tools \ required \ to \ process \ a \ job.$$

The values of these parameters are given in Table 5.23 for different instance sizes.

For each test of dimension (M, N), 10 matrices are randomly generated. For each job j, the corresponding column of A is generated as follows: an integer t_j is generated with uniform distribution in the interval $[min, max]$; t_j denotes the number of tools required by job j. Successively, a set of integers is generated uniformly in the interval $[0, M]$: these numbers identify the tools required by j. Moreover, during the process the generation of a new column

Table 5.23. *min-max* intervals of each test instance

N	M	min	max
10	10	2	4
20	15	2	6
40	30	5	15
60	40	7	20

dominating an existing one has been avoided: when this occurred we made a further random generation. This generation cannot prevent the existence of columns with all values equal to 0. In our experience, only two matrices contained columns will all zeros (two matrices of size $(20, 15)$, with one and three columns of zeros, respectively). Each instance of type (M, N, C) is obtained by means of a combination of a tool/job matrix and one of the four capacities C_1, C_2, C_3 and C_4 provided in Table 5.24.

Table 5.24. The different capacities used in the test instances

N	M	C_1	C_2	C_3	C_4
10	10	4	5	6	7
20	15	6	8	10	12 ·
40	30	15	17	20	25
60	40	20	22	25	30

The matrices utilized in this section are the same as those utilized by Crama in [58] and are available at the site [167]. Recalling that no exact solutions are available in the literature, our analysis is performed along the lines:

1. Overall ability of the presented techniques compared with the best results presented in [58].
2. Specific evaluation of each presented algorithm with solutions in [58].
3. Algorithm evaluation on each instance type.

Overall Evaluation

Let us denote with $best(I)$ the best among the solutions presented in [58] on instance I. In Table 5.25, we give the average values of $best(I)$ for different combinations of the magazine capacity and the size of the test problems.

Table 5.25. Average values of $best(I)$

N M	C_1	C_2	C_3	C_4
10 10	13.2	11.2	10.3	10.1
20 15	26.5	21.6	20.0	19.6
40 30	113.6	95.9	76,8	56.8
60 40	211.6	189.7	160.5	127.4

It should be noted that we tested all the algorithms proposed on each instance and selected the best solutions among those so obtained. Thus, these values are averaged on the 10 realizations of each problem type obtaining the value denoted as $H(I)$. In Tables 5.26 and 5.27, we give the best solutions obtained and the average values with and without extra capacity. This latter aspect has been introduced to determine the number of setups and the relative gap (δ) between $H(I)$ and $best(I)$ presented in [58]. Such a gap is computed as:

$$\delta(I) = (H(I) - best(I))/best(I) * 100.$$

Table 5.26. Best results achieved by the four algorithms on the 160 test instances

Instance	$(N * M) = (10 * 10)$					$(N * M) = (15 * 20)$			
	$C = 4$	5	6	7		$C = 6$	8	10	12
1	8	6	4	3	1	23	18	12	8
2	12	9	6	4	2	20	15	11	8
3	11	8	4	3	3	22	16	13	9
4	10	7	5	3	4	22	16	13	9
5	9	6	4	3	5	23	17	13	8
6	10	7	4	3	6	24	20	14	10
7	8	7	5	3	7	21	17	11	8
8	12	8	6	4	8	27	19	13	9
9	9	7	5	3	9	19	12	8	6
10	9	6	4	3	10	20	15	10	8
Sum	98	71	47	32	Sum	221	165	118	83
Average+C	13.8	12.1	10.7	10.2	Average+C	28.1	24.5	21.8	20.3
Best	13.2	11.2	10.3	10.1	Best	26.5	21.6	20.0	19.6
Gap	4.55	8.04	3.88	0.99	Gap	6.04	13.43	9.00	3.57

Table 5.27. Best results achieved by the four algorithms on the 160 test instances

Instance	$(N * M) = (30 * 40)$					$(N * M) = (40 * 60)$			
	$C = 15$	17	20	25		$C = 20$	22	25	30
1	112	95	73	43	1	215	184	161	125
2	111	85	68	43	2	225	199	168	128
3	98	80	62	39	3	207	183	163	120
4	108	90	71	45	4	214	192	159	120
5	116	98	76	49	5	218	195	164	121
6	99	77	68	41	6	219	193	159	116
7	115	93	73	42	7	225	197	164	121
8	126	102	82	51	8	234	204	170	127
9	103	88	67	42	9	199	179	153	115
10	106	91	69	43	10	194	175	155	121
Sum	1094	899	709	438	Sum	2150	1901	1616	1214
Average+C	124.4	106.9	90.9	68.8	Average+C	235.0	212.1	186.6	151.4
Best	113.6	95.9	76.8	56.8	Best	211.6	189.7	160.5	127.4
Gap	9.51	11.47	18.36	21.13	Gap	11.06	11.81	16.26	18.84

Performance Evaluation of Each Technique

In this section, we compare the results provided by each technique proposed with those presented in [58]. The measure used is $\delta(I)$.

This gap is computed in [58] for every instance and is averaged for the problems of the same type. The tests performed are discussed separately for each algorithm.

Algorithm 1

The analysis of Algorithm 1 performed on the 160 test instances leads to the results shown in Tables 5.28 and 5.29.

In Table 5.30, we compare Algorithm 1 with the techniques described in [58]. The first column reports the instance type. The second column gives the best results achieved by Algorithm 1. The other columns give results obtained by the algorithms in [58]. The generic entry in the table is the percentage gap between the solution of each algorithm and the value $best(I)$ shown in Tables 5.26-5.26. The symbol * denotes the case where our algorithm obtains a worse solution.

Algorithm 2

The analysis of Algorithm 2 performed on the 160 test instances leads to the results reported in Tables 5.31 and 5.32.

Table 5.33 for Algorithm 2 is the same as Table 5.30 for Algorithm 1.

Table 5.28. Results of Algorithm 1 on the 160 test instances

Instance	$(N * M) = (10 * 10)$					$(N * M) = (15 * 20)$			
	$C = 4$	5	6	7		$C = 6$	8	10	12
1	9	6	4	3	1	24	19	16	13
2	12	9	8	4	2	26	20	14	11
3	11	9	5	5	3	27	22	19	11
4	10	9	6	5	4	27	22	19	11
5	9	8	5	5	5	26	20	15	12
6	11	8	5	3	6	29	21	14	13
7	9	7	6	4	7	23	20	13	11
8	13	8	8	5	8	29	22	17	12
9	10	8	6	4	9	19	13	13	8
10	11	8	7	4	10	26	18	13	10
Sum	105	80	60	42	Sum	230	197	140	112
Average+C	14.5	13	12	11.2	Average+C	29	27.7	24	23.2
Best	13.2	11.2	10.3	10.1	Best	26.5	21.6	20.0	19.6
Gap	9.85	16.07	16.50	10.89	Gap	9.43	28.24	20.00	18.37

Table 5.29. Results of Algorithm 1 on the 160 test instances

Instance	$(N * M) = (30 * 40)$					$(N * M) = (40 * 60)$			
	$C = 15$	17	20	25		$C = 20$	22	25	30
1	123	107	86	54	1	225	201	190	140
2	115	94	80	46	2	247	209	192	150
3	107	103	79	50	3	226	198	190	143
4	116	105	85	50	4	234	205	194	139
5	129	106	88	56	5	229	209	188	154
6	99	96	77	50	6	237	218	203	145
7	115	108	84	56	7	237	204	195	141
8	130	115	96	58	8	236	218	196	143
9	110	99	71	52	9	222	201	191	142
10	107	98	79	50	10	217	192	178	130
Sum	1151	1031	825	522	Sum	2310	2055	1917	1427
Average+C	130.1	120.1	102.5	77.2	Average+C	251	227.5	216.7	172.7
Best	113.6	95.9	76.8	56.8	Best	211.6	189.7	160.5	127.4
Gap	14.52	25.23	33.46	35.92	Gap	18.62	19.93	35.02	35.56

Table 5.30. Comparison between Algorithm 1 and the algorithms given in [58]

	Algo 1	1	2	3	4	5	6	7	8	9
10/10/04	9.85	12.1	14.3	12.3	* 4.6	22.6	26.0	* 8.7	* 5.8	41.2
10/10/05	16.07	19.0	* 13.6	* 8.1	* 3.7	* 14.1	24.3	* 7.4	* 10.1	33.8
10/10/06	16.5	17.8	* 9.7	* 5.7	* 2.9	* 9.7	18.3	* 3.0	* 6.7	26.3
10/10/07	10.89	11.7	* 3.9	* 1.0	0.0	* 3.0	* 9.8	* 0.0	* 3.0	13.8
15/20/06	9.43	15.5	12.0	13.7	* 4.6	25.7	33.6	10.0	12.3	45.9
15/20/08	28.24	37.3	* 13.9	* 11	* 4.6	* 20.4	35.7	* 9.7	* 23.8	42.2
15/20/10	20	30.5	* 8.3	* 5.6	* 1.5	* 10.4	24.3	* 6.4	25.6	30.1
15/20/12	18.37	* 15.3	* 2.1	* 1.0	* 0	* 3.5	* 13.6	* 1.0	* 16.6	* 18.1
40/30/15	14.52	* 9.4	* 8.8	* 11.4	* 6.2	30.5	30.3	* 6.0	16.6	42.9
40/30/17	25.23	* 16.3	* 9.4	* 9.8	* 5.5	31.2	31.0	* 4.5	27.5	44.6
40/30/20	33.46	33.8	* 12.1	* 9.8	* 3.2	* 30.4	* 33.0	* 6.0	35.1	45.5
40/30/25	35.92	39.4	* 15.0	* 8.3	* 2.6	* 27.8	* 34.5	* 6.1	37.8	40.5
40/60/20	18.62	* 6.9	* 9.7	* 10.2	* 5.8	30.6	25.8	* 4.8	20.0	37.1
40/60/22	19.93	* 9.9	* 8.7	* 7.9	* 3.3	29.3	25.4	* 3.7	25.4	36.5
40/60/25	35.02	* 21.8	* 10.5	* 8.2	* 2.8	* 30.2	* 29.7	* 2.1	35.5	38.0
40/60/30	35.56	36.7	* 13.1	* 6.5	* 1.7	* 28.8	* 30.1	* 4.5	36.7	37.6

Table 5.31. Results of Algorithm 2 on the 160 test instances

	$(N * M) = (10 * 10)$					$(N * M) = (15 * 20)$			
Instance	$C = 4$	5	6	7		$C = 6$	8	10	12
1	8	8	5	3	1	23	19	12	9
2	13	9	6	4	2	20	15	13	8
3	11	8	4	3	3	25	17	15	9
4	12	7	5	3	4	25	17	15	9
5	9	6	5	3	5	23	17	13	8
6	10	8	4	4	6	26	20	16	10
7	10	7	5	3	7	21	17	11	8
8	12	8	6	4	8	27	20	15	10
9	9	7	5	3	9	19	13	8	6
10	10	8	5	4	10	20	16	12	9
Sum	104	76	50	34	Sum	229	171	130	86
Average+C	14.4	12.6	11	10.4	Average+C	28.9	25.1	23	20.6
Best	13.2	11.2	10.3	10.1	Best	26.5	21.6	20.0	19.6
Gap	9.09	12.50	6.80	2.97	Gap	9.06	16.20	15.00	5.10

Table 5.32. Results of Algorithm 2 on the 160 test instances

Instance	$C = 15$	17	20	25		Instance	$C = 20$	22	25	30
	$(N * M) = (30 * 40)$						$(N * M) = (40 * 60)$			
1	112	95	75	45		1	215	184	161	130
2	111	85	68	50		2	225	199	173	138
3	98	82	67	44		3	207	183	170	130
4	108	90	71	51		4	214	196	175	135
5	116	100	76	58		5	218	195	164	134
6	99	78	69	44		6	219	198	179	130
7	115	93	78	50		7	225	198	171	140
8	126	102	88	53		8	235	204	174	147
9	103	88	68	47		9	199	179	162	133
10	106	91	70	50		10	196	176	155	121
Sum	1094	904	730	492		Sum	2153	1912	1684	1338
Average+C	124.4	107.4	93.0	74.2		Average+C	235.3	213.2	193.4	163.8
Best	113.6	95.9	76.8	56.8		Best	211.6	189.7	160.5	127.4
Gap	9.51	11.99	21.09	30.63		Gap	11.20	12.39	20.50	28.57

Table 5.33. Comparison between Algorithm 2 and the algorithms given in [58]

	Algo 2	1	2	3	4	5	6	7	8	9
10/10/04	9.09	12.1	14.3	12.3	* 4.6	22.6	26.0	* 8.7	* 5.8	41.2
10/10/05	12.5	19.0	13.6	* 8.1	* 3.7	14.1	24.3	* 7.4	* 10.1	33.8
10/10/06	6.8	17.8	9.7	* 5.7	* 2.9	9.7	18.3	* 3.0	* 6.7	26.3
10/10/07	2.97	11.7	3.9	* 1.0	* 0.0	3	9.8	* 0.0	3.0	13.8
15/20/06	9.06	15.5	12	13.7	* 4.6	25.7	33.6	* 10.0	12.3	45.9
15/20/08	16.2	37.3	* 13.9	* 11.0	* 4.6	20.4	35.7	* 9.7	23.8	42.2
15/20/10	15.0	30.5	* 8.3	* 5.6	* 1.5	* 10.4	24.3	* 6.4	25.6	30.1
15/20/12	5.1	15.3	* 2.1	* 1	* 0.0	* 3.5	13.6	* 1.0	16.6	18.1
40/30/15	9.51	* 9.4	* 8.8	11.4	* 6.2	30.5	30.3	* 6.0	16.6	42.9
40/30/17	11.99	16.3	* 9.4	* 9.8	* 5.5	31.2	31.0	* 4.5	27.5	44.6
40/30/20	21.09	33.8	* 12.1	* 9.8	* 3.2	30.4	33.0	* 6.0	35.1	45.5
40/30/25	30.63	39.4	* 15	* 8.3	* 2.6	* 27.8	34.5	* 6.1	37.8	40.5
40/60/20	11.2	* 6.9	* 9.7	* 10.2	* 5.8	30.6	25.8	* 4.8	20.0	37.1
40/60/22	12.39	* 9.9	* 8.7	* 7.9	* 3.3	29.3	25.4	* 3.7	25.4	36.5
40/60/25	20.5	21.8	* 10.5	* 8.2	* 2.8	30.2	29.7	* 2.1	35.5	38.0
40/60/30	28.57	36.7	* 13.1	* 6.5	* 1.7	28.8	30.1	* 4.5	36.7	37.6

Algorithm 3

The analysis of Algorithm 3 performed on the 160 test instances leads to the results reported in Tables 5.34 and 5.35.

Table 5.34. Results of Algorithm 3 on the 160 test instances

	$(N * M) = (10 * 10)$					$(N * M) = (15 * 20)$			
Instance	$C = 4$	5	6	7		$C = 6$	8	10	12
1	8	8	5	3	1	23	19	12	9
2	13	9	6	4	2	20	17	11	8
3	12	8	4	3	3	22	16	13	9
4	11	8	5	3	4	22	16	13	9
5	10	6	5	3	5	23	17	13	8
6	10	8	4	4	6	24	20	15	11
7	8	7	5	3	7	22	17	13	8
8	12	8	6	4	8	27	20	15	11
9	9	7	5	3	9	20	12	8	6
10	9	8	5	4	10	20	16	12	9
Sum	102	77	50	34	Sum	223	170	125	88
Average+C	14.2	12.7	11.0	10.4	Average+C	28.3	25.0	22.5	20.8
Best	13.2	11.2	10.3	10.1	Best	26.5	21.6	20	19.6
Gap	7.58	13.39	6.80	2.97	Gap	6.79	15.74	12.50	6.12

Table 5.35. Results of Algorithm 3 on the 160 test instances

	$(N * M) = (30 * 40)$					$(N * M) = (40 * 60)$			
Instance	$C = 15$	17	20	25		$C = 20$	22	25	30
1	113	96	75	43	1	219	186	167	131
2	114	85	69	50	2	228	203	171	143
3	103	85	66	45	3	213	186	168	130
4	110	92	74	51	4	226	198	176	139
5	116	98	77	58	5	224	199	166	134
6	99	77	68	43	6	223	207	188	133
7	116	95	77	50	7	225	199	181	141
8	128	103	89	54	8	234	210	177	147
9	107	88	69	47	9	203	185	165	135
10	106	93	69	48	10	194	175	156	121
Sum	1112	912	733	489	Sum	2189	1948	1715	1354
Average+C	126.2	108.2	93.3	73.9	Average+C	238.9	216.8	196.5	165.4
Best	113.6	95.9	76.8	56.8	Best	211.6	189.7	160.5	127.4
Gap	11.09	12.83	21.48	30.11	Gap	12.90	14.29	22.43	29.83

Table 5.36 for Algorithm 3 is the same as Tables 5.30 for Algorithm 1 and 5.33 for Algorithm 2.

Table 5.36. Comparison between Algorithm 3 and the algorithms given in [58]

	Algo 3	1	2	3	4	5	6	7	8	9
10/10/04	7.58	12.1	14.3	12.3	* 4.6	22.6	26	8.7	* 5.8	41.2
10/10/05	13.39	19	13.6	* 8.1	* 3.7	14.1	24.3	* 7.4	* 10.1	33.8
10/10/06	6.8	17.8	9.7	* 5.7	* 2.9	9.7	18.3	* 3	* 6.7	26.3
10/10/07	2.97	11.7	3.9	* 1	* 0	3	9.8	* 0	3	13.8
15/20/06	6.79	15.5	12	13.7	* 4.6	25.7	33.6	10	12.3	45.9
15/20/08	15.74	37.3	* 13.9	* 11	* 4.6	20.4	35.7	* 9.7	23.8	42.2
15/20/10	12.5	30.5	* 8.3	* 5.6	* 1.5	* 10.4	24.3	* 6.4	25.6	30.1
15/20/12	6.12	15.3	* 2.1	* 1	* 0	* 3.5	13.6	* 1	16.6	18.1
40/30/15	11.09	* 9.4	* 8.8	11.4	* 6.2	30.5	30.3	* 6	16.6	42.9
40/30/17	12.83	16.3	* 9.4	* 9.8	* 5.5	31.2	31	* 4.5	27.5	44.6
40/30/20	21.48	33.8	* 12.1	* 9.8	* 3.2	30.4	33	* 6	35.1	45.5
40/30/25	30.11	39.4	* 15	* 8.3	* 2.6	* 27.8	34.5	* 6.1	37.8	40.5
40/60/20	12.9	* 6.9	* 9.7	* 10.2	* 5.8	30.6	25.8	* 4.8	20	37.1
40/60/22	14.29	* 9.9	* 8.7	* 7.9	* 3.3	29.3	25.4	* 3.7	25.4	36.5
40/60/25	22.43	* 21.8	* 10.5	* 8.2	* 2.8	30.2	29.7	* 2.1	35.5	38
40/60/30	29.83	36.7	* 13.1	* 6.5	* 1.7	* 28.8	30.1	* 4.5	36.7	37.6

Algorithm 4

The analysis of Algorithm 4 performed on the 160 test instances leads to the results given in Tables 5.37 and 5.38.

Table 5.39 for Algorithm 4 is the same as Tables 5.30 for Algorithm 1, 5.33 for Algorithm 2, and 5.39 for Algorithm 4.

5.4.3 Comparison Among the Proposed Algorithms

Similarly to what was done in [58], we describe a comparison among the proposed heuristics, always taking the average values over the 10 tests (see Table 5.40).

Organizing data in a histogram as shown in Figure 5.20 we observe a reasonable equivalence for small dimensions. As soon as the size increases,

Table 5.37. Results of Algorithm 4 on the 160 test instances

Instance	$(N*M) = (10*10)$					$(N*M) = (15*20)$			
Instance	$C=4$	5	6	7		$C=6$	8	10	12
1	9	6	4	3	1	25	18	12	8
2	12	9	6	4	2	22	15	11	8
3	13	9	7	5	3	27	20	15	11
4	10	7	5	3	4	27	20	15	11
5	10	6	4	3	5	27	20	16	12
6	10	7	5	3	6	28	20	15	11
7	11	8	6	4	7	25	18	14	10
8	13	10	7	5	8	28	19	13	9
9	9	7	5	3	9	22	16	12	8
10	10	6	4	3	10	23	15	10	8
Sum	107	75	53	36	Sum	254	181	133	96
Average+C	14.7	12.5	11.3	10.6	Average+C	31.4	26.1	23.3	21.6
Best	13.2	11.2	10.3	10.1	Best	26.5	21.6	20.0	19.6
Gap	11.36	11.61	9.71	4.95	Gap	18.49	20.83	16.50	10.20

Table 5.38. Results of Algorithm 4 on the 160 test instances

Instance	$(N*M) = (30*40)$					$(N*M) = (40*60)$			
Instance	$C=15$	17	20	25		$C=20$	22	25	30
1	118	96	73	45	1	217	193	164	125
2	115	94	73	43	2	230	201	168	128
3	98	80	62	39	3	222	195	163	120
4	119	97	74	45	4	220	192	159	120
5	120	99	77	49	5	226	196	165	121
6	109	90	68	41	6	219	193	159	116
7	119	97	73	42	7	225	197	164	121
8	129	105	82	51	8	238	206	170	127
9	109	89	67	42	9	205	182	153	115
10	113	91	70	43	10	201	180	156	122
Sum	1149	938	719	440	Sum	2203	1935	1621	1215
Average+C	129.9	110.8	91.9	69.0	Average+C	240.3	215.5	187.1	151.5
Best	113.6	95.9	76.8	56.8	Best	211.6	189.7	160.5	127.4
Gap	14.35	15.54	19.66	21.48	Gap	13.56	13.60	16.57	18.92

Algorithm 2 performs better than the others. Only very large instances allow Algorithm 4 to do better.

Now we examine the ratio max/C. We call the matrices for which such a ratio is small "sparse", and those for which this ratio is high "dense". For the instances (dense matrices)

$$(10, 10, 4)\ (20, 15, 6)\ (30, 40, 15)\ (60, 40, 20)$$

Table 5.39. Comparison between Algorithm 4 and the algorithms given in [58]

	Algo 4	1	2	3	4	5	6	7	8	9
10/10/04	11.36	12.1	14.3	12.3	* 4.6	22.6	26	* 8.7	* 5.8	41.2
10/10/05	11.61	19.0	13.6	* 8.1	* 3.7	14.1	24.3	* 7.4	* 10.1	33.8
10/10/06	9.71	17.8	* 9.7	* 5.7	* 2.9	* 9.7	18.3	* 3.0	* 6.7	26.3
10/10/07	4.95	11.7	*3.9	* 1.0	* 0.0	* 3.0	9.8	* 0.0	* 3.0	13.8
15/20/06	18.49	* 15.5	* 12.0	* 13.7	* 4.6	25.7	33.6	* 10.0	* 12.3	45.9
15/20/08	20.83	37.3	* 13.9	* 11.0	* 4.6	* 20.4	35.7	* 9.7	23.8	42.2
15/20/10	16.50	30.5	* 8.3	* 5.6	* 1.5	* 10.4	24.3	* 6.4	25.6	30.1
15/20/12	10.2	15.3	* 2.1	* 1.0	* 0.0	* 3.5	13.6	* 1.0	16.6	18.1
40/30/15	14.35	* 9.4	* 8.8	11.4	* 6.2	30.5	30.3	* 6.0	16.6	42.9
40/30/17	15.54	16.3	* 9.4	* 9.8	* 5.5	31.2	31.0	* 4.5	27.5	44.6
40/30/20	19.66	33.8	* 12.1	* 9.8	* 3.2	30.4	33.0	* 6.0	35.1	45.5
40/30/25	21.48	39.4	* 15	* 8.3	* 2.6	* 27.8	34.5	* 6.1	37.8	40.5
40/60/20	13.56	* 6.9	* 9.7	* 10.2	* 5.8	30.6	25.8	* 4.8	20.0	37.1
40/60/22	13.60	* 9.9	* 8.7	* 7.9	* 3.3	29.3	25.4	* 3.7	25.4	36.5
40/60/25	16.57	21.8	* 10.5	* 8.2	* 2.8	30.2	29.7	* 2.1	35.5	38.0
40/60/30	18.92	36.7	* 13.1	* 6.5	* 1.7	28.8	30.1	* 4.5	36.7	37.6

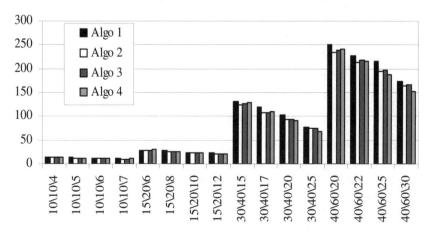

Fig. 5.20. Comparison with average values

Table 5.40. Comparison among the four algorithms proposed

	Algo 1	Algo 2	Algo 3	Algo 4
10/10/4	14.50	14.4	14.2	14.7
10/10/5	13.0	12.6	12.7	12.5
10/10/6	12.0	11.0	11.0	11.3
10/10/7	11.2	10.4	10.4	10.6
15/20/6	29.0	28.9	28.3	31.4
15/20/8	27.7	25.1	25.0	26.1
15/20/10	24.0	23.0	22.5	23.3
15/20/12	23.2	20.6	20.8	21.6
30/40/15	130.1	124.4	126.2	129.9
30/40/17	120.1	107.4	108.2	110.8
30/40/20	102.5	93.0	93.3	91.9
30/40/25	77.2	74.2	73.9	69.0
40/60/20	251.0	235.3	238.9	240.3
40/60/22	227.5	213.2	216.8	215.5
40/60/25	216.7	193.4	196.5	187.1
40/60/30	172.7	163.8	165.4	151.5

the average values are listed in Table 5.41.

Table 5.41. Average values for dense matrices

	Algo 1	Algo 2	Algo 3	Algo 4
10/10/4	14.5	14.4	14.2	14.7
15/20/6	29.0	28.9	28.3	31.4
30/40/15	130.1	124.4	126.2	129.9
40/60/20	251.0	235.3	238.9	240.3

Reporting these values on a chart, we have Figure 5.21.

It can be observed that the second algorithm has the best performance followed by the first, and then by the fourth.

For instances

$$(10, 10, 7) \ (15, 20, 12) \ (30, 40, 25) \ (40, 60, 30)$$

we get the results in Table 5.42.

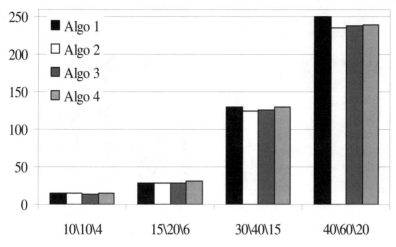

Fig. 5.21. Average values for dense matrices

Table 5.42. Average performance of the algorithms on instances 10/10/7, 15/20/12, 30/40/25, 40/60/30

	Algo 1	Algo 2	Algo 3	Algo 4
10/10/7	11.2	10.4	10.4	10.6
15/20/12	23.2	20.6	20.8	21.6
30/40/25	77.2	74.2	73.9	69.0
40/60/30	172.7	163.8	165.4	151.5

In Figure 5.22, we provide the corresponding chart.

From our analysis, we get that Algorithms 2 and 3 perform better than the ones with the grouping setup phase. The best heuristic is Algorithm 4, which guarantees smaller values as the size of the instance grows. It might be worth looking at the number of times an algorithm outperforms the others (see Table 5.43 and Figure 5.23).

Summing up, we observe that there is no dominant algorithm over all the instances. Algorithm 2 is the best among those using the grouping strategy, while Algorithm 4 is quite unstable, but seems better for larger instances.

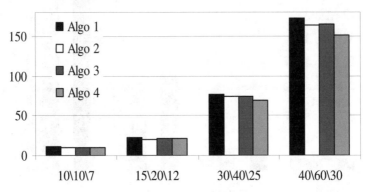

Fig. 5.22. Average performance of the algorithms on instances 10/10/7, 15/20/12, 30/40/25, 40/60/30

5.5 Conclusions

In this chapter the minimum tool changeover problem has been studied and algorithms based on different approaches have been described and computa-

Table 5.43. Number of times an algorithm outperforms the others

10/10/4	4	6	6	4
10/10/5	4	7	6	7
10/10/6	1	7	7	6
10/10/7	3	8	8	7
15/20/6	1	7	8	0
15/20/8	0	4	6	5
15/20/10	1	4	6	4
15/20/12	0	7	6	4
30/40/15	2	10	3	1
30/40/17	0	7	4	2
30/40/20	0	3	2	6
30/40/25	3	4	5	9
40/60/20	0	8	3	2
40/60/22	0	6	1	3
40/60/25	0	3	0	7
40/60/30	0	1	1	9

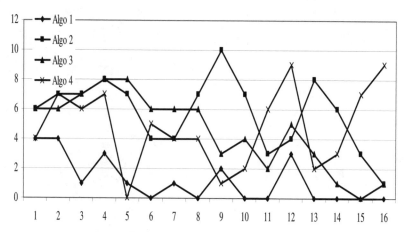

Fig. 5.23. Number of times each algorithm outperforms the others

tionally evaluated. Two families of heuristics have been designed. The first uses first a grouping approach and then a sequencing one. Jobs are grouped together if they require C tools. They are then sequenced according to a strategy similar to KTNS. The second approach does not group, but considers only compatibilities on successive jobs modelled by means of a specific graph model.

A computational comparison is made with respect to the heuristics of Crama *et al.* and Tang and Denardo [58, 164], indexCrama performing better than Tang and Denardo [164] in all cases and, in several cases, also better than Crama *et al.* [58].

Appendix A

Implementation Code of Chapter 3

```
#include<io.h>
#include<stdio.h>
#include<stdlib.h>
#include<time.h>
#include<math.h>

#define maxint 10000000
#define graph_nodes 100
#define maxiteration graph_nodes*100000
#define density 8
#define max_resource 4
#define max_quantity 10
#define alpha 10
#define aa 135

int adj[graph_nodes][graph_nodes];
int p[graph_nodes], t[graph_nodes], update[graph_nodes];
double delta[graph_nodes];
int T, Tstar, graph_arcs;
int i, j, k;
int iterations;
int ass[graph_nodes], order[graph_nodes];
int stop, count, maxi;
double old_penalty, current_penalty, penalty,penaltystar;
int checkpoint, check;
int total_R[graph_nodes][max_resource];
```

```c
int R[graph_nodes][max_resource]; /*Quantity of a resource requested by a
task*/
int max[max_resource];
int num_scheduled[graph_nodes];
double penaltyTstar;
int Tpenaltystar;
int Tmax;
int greedy;

FILE *fd;

time_t starting,ending;

void Greedy_Scheduler(void);
double Penalty(void);

void Update_T(void)
{
  register int i;

  current_penalty=Penalty();
  T=0;
  for (i=0; i<graph_nodes;i++)
       if (t[i]>T)
  T=t[i];
  if(T>Tmax){
    Tmax=T;
    printf("Tmax: %i",Tmax);
  }
  if (T<Tstar){
    Tstar=T;
    penaltyTstar=current_penalty;
    ending=time(NULL);
    printf("Tstar: %i",Tstar);
    printf("Best time slots: %d, penalty: %0.2f, iter.: %i, sec.: %0.2f, peak:
%i ",Tstar,penaltyTstar,iterations,difftime(ending,starting),maxi);
    printf("%i %i",Tstar,iterations);
  }
  if (T==Tstar)
    if (penaltyTstar>current_penalty){
      penaltyTstar=current_penalty;
```

```c
    ending=time(NULL);
    printf("Best time slots: %d, penalty: %0.2f, iter.: %i, sec.: %0.2f,
    peak: %i",Tstar,penaltyTstar,iterations,difftime(ending,starting),maxi);
    printf("%i",Tstar);
    }
}


void Update_Penalty(void)
{
  current_penalty=Penalty();
  if (current_penalty<penaltystar){
    penaltystar=current_penalty;
    Tpenaltystar=T;
    ending=time(NULL);
    printf("Best penalty: %0.2f, time slots: %d, iter.: %i, sec.: %0.2f, peak:
    %i",penaltystar,Tpenaltystar,iterations,difftime(ending,starting),maxi);
    printf("%0.2f %i",penaltystar,iterations);
    }
  if (current_penalty==penaltystar)
    if (T<Tpenaltystar){
      Tpenaltystar=T;
      ending=time(NULL);
      printf("Best penalty: %0.2f, time slots: %d, iter.: %i, sec.: %0.2f,
      peak: %i",penaltystar,Tpenaltystar,iterations,difftime(ending,starting),
      maxi);
      printf("%i",Tpenaltystar);
    }
}


void Stopping_rule(void)
{
  if (iterations>maxiteration){
    ending=time(NULL);
    printf("CPU time: %f seconds",difftime(ending,starting));
    printf("Best number of time slots found is %d with penalty %f",Tstar,
    penaltyTstar);
    printf("Best penalty is %f with time slots %d",penaltystar,Tpenaltystar);
    exit(1);
  }
}
```

```
void Order_set_decreasing_d(double par[graph_nodes], int set[graph_nodes])
{
    int i,a,b,chosen[graph_nodes];
    double max;

    for (i=0; i<graph_nodes; i++)
      chosen[i]=0;
    b=0;
    while (b<graph_nodes) {
      max=0.00;
      for (i=0; i<graph_nodes; i++)
        if (chosen[i]==0)
          if (par[i]>=max){
            max=par[i];
            a=i;
          }
      chosen[a]=1;
      set[b]=a;
      b++;
    }
}


void Order_set_decreasing_i(int par[graph_nodes], int set[graph_nodes])
{
    int i,a,b,chosen[graph_nodes],max;

    for (i=0; i<graph_nodes; i++)
      chosen[i]=0;
    b=0;
    while (b<graph_nodes) {
      max=0;
      for (i=0; i<graph_nodes; i++)
        if (chosen[i]==0)
          if (par[i]>=max){
            max=par[i];
            a=i;
          }
      chosen[a]=1;
      set[b]=a;
      b++;
    }
```

```
}

void Order_set_increasing(int par[graph_nodes], int set[graph_nodes])
{
   int i,a,b,chosen[graph_nodes],min;

   for (i=0; i<graph_nodes; i++)
     chosen[i]=0;
   b=0;
   while (b<graph_nodes) {
     min=maxint;
     for (i=0; i<graph_nodes; i++)
       if (chosen[i]==0)
         if (par[i]<=min){
           min=par[i];
           a=i;
         }
     chosen[a]=1;
     set[b]=a;
     b++;
     }
}

void Checkpointing(void)
{
   Stopping_rule();
   check=check+checkpoint;
   Order_set_increasing(update,order);
   for (i=0; i<graph_nodes; i++){
     update[i]=0;
     t[i]=graph_nodes;
   }
   Greedy_Scheduler();
}

void Assign_Lowest(int a) {
   register int i;

   for (i=0; i<graph_nodes; i++)
     ass[i]=1;
   for (i=0; i<graph_nodes; i++)
```

```
    if (adj[i][a]==1)
      ass[t[i]-1]=0;
  for (i=0; i<graph_nodes; i++)
    if (ass[i]==1) {
      t[a]=i+1;
      break;
    }
  iterations++;
  update[a]++;
  if (iterations>=check)
    Checkpointing();
}

void Set_Checkpoint(void)
{
  checkpoint=alpha*graph_nodes;
  check=checkpoint;
}

void Order_Nodes_WRT_Degree(void)
{
  int degree[graph_nodes];

  for (i=0; i<graph_nodes; i++) {
    count=0;
    for (j=0; j<graph_nodes; j++)
      if (adj[i][j]==1)
    count++;
    degree[i]=count;
  }
  for (i=0; i<graph_nodes; i++)
    p[i]=degree[i];
}

int Admissible(int a, int b)
{
  int i;

  for (i=0; i<graph_nodes; i++)
    if (adj[i][a]==1)
      if (t[i]==b)
```

```
      return 0;
   return 1;
}

void Total_R(void)
{
   int i,j,k;

   for (i=0; i<T; i++)
     for (j=0; j<max_resource; j++)
       total_R[i][j]=0;
   for (i=0; i<T; i++)
     num_scheduled[i]=0;
     for (j=0; j<graph_nodes; j++){
       if(t[j]==(i+1)){
         for(k=0; k<max_resource; k++)
           total_R[i][k]=total_R[i][k]+R[j][k];
         num_scheduled[i]++;
       }
     }
   }
}

void Calculate_Max(void)
{
   int i,j;

   for (i=0; i<max_resource; i++)
     max[i]=0;
   for (i=0; i<T; i++)
     for (j=0; j<max_resource; j++)
       if (max[j]<total_R[i][j])
         max[j]=total_R[i][j];
}

double Deviation(void)
{
   int i,sum,count,real_num_resource;
   double mean,abs_deviation;
   sum=0;
   count=0;
```

```
  for (i=0; i<max_resource; i++){
    sum=sum+max[i];
    if (max[i]==0)
      count++;
  }
  real_num_resource=max_resource-count;
  mean=(sum+0.00)/(real_num_resource+0.00);
  abs_deviation=0.00;
  maxi=0;
  for (i=0; i<max_resource; i++)
    if (max[i]!=0){
      if(max[i]-mean¿0)
        abs_deviation=abs_deviation+max[i]-mean;
    else
      abs_deviation=abs_deviation+mean-max[i];
    if (maxi¡max[i])
      maxi=max[i];
}
  return abs_deviation;
}

double Penalty(void)
{
  Total_R();
  Calculate_Max();
  penalty=Deviation();
  return penalty;
}

void Delta(int a, int b)
{
  int old_slot;

  old_slot=t[a];
  t[a]=b;
  delta[a]=Penalty()-old_penalty;
  t[a]=old_slot;
}

void Penalty_Trader(void)
{
```

```
  int chosen[graph_nodes];

  old_penalty=Penalty();
  Update_T();
  for (i=0; i<graph_nodes; i++)
    Delta(i,T+1);
  Order_set_decreasing_d(delta,order);
  for (i=0; i<graph_nodes; i++)
    chosen[i]=0;
  for (i=0; i<graph_nodes; i++)
    if (delta[order[i]]>=0){
      count=0;
  for (j=0; j<graph_nodes; j++)
    if (chosen[j]==1)
      if (adj[order[i]][j]==1)
        count++;
      if (count==0)
        update[order[i]]++;
      t[order[i]]=T+1;
      chosen[order[i]]=1;

  iterations++;
  if (iterations>=check)
    Checkpointing();
  }
  Update_T();
  Update_Penalty();
  current_penalty=Penalty();
  printf("Penalty: %f and Time Slots %i ",current_penalty,T);
  if (old_penalty==current_penalty)
    Penalty_Trader();
  else {
    Order_set_increasing(t,order);
    Greedy_Scheduler();
  }
}

void Try_to_change_slot(int a)
{
  int j,old_slot;
  double k;
```

```
  current_penalty=old_penalty;
  for(j=0; j<T; j++)
    if(t[a]!=(j+1))
      if (Admissible(a,j+1)==1){
        old_slot=t[a];
        t[a]=j+1;
        k=Penalty();
        if(k<=current_penalty){
          current_penalty=k;
          update[a]++;
        }
      else
    t[a]=old_slot;
    }
  iterations++;
  if (iterations>=check)
    Checkpointing();
}

void Order_Num_Scheduled(void)
{
  int chosen[graph_nodes],chosen2[graph_nodes];
  int max,max2,max3,a,b,c,d,e,i;

  for (i=0; i<graph_nodes; i++){
    chosen[i]=0;
    chosen2[i]=0;
  }
  d=0;
  while (d<graph_nodes) {
    max=0;
    a=-1;
    for (i=0; i<T; i++)
      if ((chosen[i]==0)&&(num_scheduled[i]!=0))
        if (num_scheduled[i]>=max){
          max=num_scheduled[i];
          a=i; /*index of the time slot with the highest num_scheduled*/
        }
    if (a!=-1){
      chosen[a]=1;
```

```
    max2=0;
    for (i=0; i<max_resource; i++)
      if(total_R[a][i]>max2){
        max2=total_R[a][i];
        b=i; /*index of resource with highest sum requirement*/
      }
    e=0;
    while (e<num_scheduled[a]){
      max3=0;
      for (i=0; i<graph_nodes; i++)
        if((t[i]==(a+1))&&(chosen2[i]==0))
          if (R[i][b]>=max3){
            max3=R[i][b];
            c=i; /*activity*/
          }
      chosen2[c]=1;
      order[d]=c;
      d++;
      e++;
    }
  }
 }
}

void Penalty_Decreaser(void)
{
  Order_Num_Scheduled();
  old_penalty=Penalty();
  for (i=0; i<graph_nodes; i++)
    Try_to_change_slot(order[i]);
  current_penalty=Penalty();
  Update_T();
  Update_Penalty();
  if (old_penalty<=current_penalty)
    Penalty_Trader();
  else
    Penalty_Decreaser();
}

void Greedy_Scheduler(void)
{
```

```c
  int k;

  greedy++;
  printf("GD: %i It: %i",greedy,iterations);
  k=0;
  while (k<graph_nodes) {
    Assign_Lowest(order[k]);
    k++;
  }
  Update_T();
  Penalty();
  Update_Penalty();
  Penalty_Decreaser();
}

void Build_Graph(void)
{
  int i,j;

  graph_arcs=0;
  for(i=0; i<graph_nodes-1; i++)
    for(j=i+1; j<graph_nodes; j++)
      if(rand()%10<density){
        adj[i][j]=1;
        adj[j][i]=1;
        graph_arcs++;
      }
      else {
        adj[i][j]=0;
        adj[j][i]=0;
      }
  for(i=0; i<graph_nodes; i++)
    adj[i][i]=0;
  fd=fopen("file_name.txt","w");
  fprintf(fd,"%i %i",graph_nodes,graph_arcs);
  for(i=0; i<graph_nodes; i++)
    for(j=i+1; j<graph_nodes; j++)
      if(adj[i][j]==1)
  fprintf(fd,"e %i %i",i,j);
  fclose(fd);
}
```

```
void Associate_Resources(void)
{
   int i,j,count;
   float sum,min,max;

   for(i=0; i<graph_nodes; i++){
     for(j=0; j<max_resource; j++)
       if (rand()%10<5)
         R[i][j]=rand()%max_quantity+1;
       else
         R[i][j]=0;
       count=0;
       for(j=0; j<max_resource; j++)
         if(R[i][j]!=0)
       count++;
         if (count==0)
       R[i][rand()%max_resource+1]=rand()%max_quantity+1;
   }
   min=1000.0;
   max=0.0;
   for(j=0; j<max_resource; j++){
     sum=0.0;
   for(i=0; i<graph_nodes; i++)
     sum=sum+R[i][j];
   sum=sum/aa;
   if(sum<min)
     min=sum;
   if(sum>max)
     max=sum;
   }
   printf("min: %f, max: %f",min,max);
   getchar();
}

void Initialize(void)
{
   int i;

   for (i=0; i<graph_nodes; i++){
     t[i]=graph_nodes;
```

```
      update[i]++;
      max[i]=0;
    }
    greedy=0;
    Tmax=0;
    T=graph_nodes;
    Tstar=graph_nodes;
    Tpenaltystar=graph_nodes;
    penalty=maxint+0.00;
    penaltystar=maxint+0.00;
    penaltyTstar=maxint+0.00;
    iterations=0;
    Set_Checkpoint();
    Order_Nodes_WRT_Degree();
    Order_set_decreasing_i(p,order);
    Greedy_Scheduler();
}

void main(int argc, char *argv[])
{
    Build_Graph();
    Associate_Resources();
    starting=time(NULL);
    Initialize();
}
```

Implementation Code of Chapter 1.15

```
#include<io.h>
#include<stdlib.h>
#include<stdio.h>
#define n 40 /*number of jobs*/
#define num_mac 7 /*number of machines*/
#define pr 2 /*pr is the probability that A6 gives 1 on load*/

int i, ml, t, j;
int A1[3*n], A2[3*n], A3[num_mac], A5[3*n], A6[3*n];
int AssMacTask[3*n][num_mac];
int A[3*n]; /*In vector A Load corresponds to 1, Make to 2, and Unload to
3; 4 means that the task has been processed*/
int A2ini[3*n]; /*It stores the time from which, when the time is activated,
the idle time is calculated*/
int end[3*n]; /*It stores the instant of time when the operation will be fin-
ished*/

int Verify(int a) /*It determines if task a is a Make operation*/
{
    int c;
    c=0;
    while(c<n){
      if(a==(1+3*c))
        return 0;
      else
        c++;
    }
    return 1;
}

int Free_Machine(int a, int b) /*Determine if at a certain time slot a ma-
chine b is free*/
{
    int z;

    for (z=0; z<num_mac; z++)
      if (AssMacTask[b][z]==1)
        if (a>=A3[z]){
          ml=z;
          return 1;
```

```
      }
   return 0;
}

int Find_max(int a)
{
   int max;
   int max_k;
   int k;
   int save_max,save_max_k;

   max=-1;
   for (k=0; k<3*n; k++){
     if ((A2[k]>=max)&&(A[k]!=4)&&Free_Machine(a,k)&&(A5[k]==1)
     &&(A[k]!=2)){
        save_max=max;
        max=A2[k];
        save_max_k=max_k;
        max_k=k;
     }
     if((A[k]==1)&&(!Free_Machine(a,k+1))){
        max=save_max;
        max_k=save_max_k;
     }
   }
   if (max==-1)
     return 4*n;
   else
     return max_k;
}

void Instance_Generation(void)
{
   int z2;
   int x;
   int i2,i4;
   int assign;

   x=1;
   for (i2=0; i2<3*n; i2++){
     if (x==1){
```

```
    A[i2]=1;
    x=2;
  }
  else
    if (x==2){
      A[i2]=2;
      x=3;
    }
    else
      if (x==3){
        A[i2]=3;
        x=1;
      }
}
for (i=0; i<3*n; i++){
  A3[i]=0;
  end[i]=0;
}
for (i=0; i<3*n; i++)
  if (Verify(i)==1)
    A1[i]=rand()%20+50;
  else
    A1[i]=rand()%120+240;
x=1;
for (i2=0; i2<3*n; i2++){
  if (x==1){
    A5[i2]=1;
    A2ini[i2]=0;
    x=2;
  }
  else
    if (x==2){
      A5[i2]=0;
      x=3;
    }
    else
      if (x==3){
        A5[i2]=0;
        x=1;
      }  }
x=1;
```

```
for (i2=0; i2<3*n; i2++){
  if (x==1){
    if (rand()%10<=pr)
      A6[i2]=1;
    else
      A6[i2]=0;
    x=2;
  }
  else
    if (x==2){
      A6[i2]=0;
      x=3;
    }
  else
    if (x==3){
      A6[i2]=0;
      x=1;
    }
}
for (i2=0; i2<3*n; i2++)
  for (z2=0; z2<num_mac; z2++)
    AssMacTask[i2][z2]=0;
i4=0;
for (i2=0; i2<3*n; i2++)
  if(i2==1+3*i4){
    assign=0;
for (z2=1; z2<num_mac; z2++)
  if(assign==0)
    if (rand()%1000<=100){
      AssMacTask[i2][z2]=1;
      assign=1;
    }
  if (assign==0)
    AssMacTask[i2][rand()%(num_mac-1)+1]=1;
  i4++;
  }
  else
    AssMacTask[i2][0]=1;
}

int Finish(void)
```

```
{
  int b;

  for (b=0; b<3*n; b++)
    if (A[b]!=4)
      return 0;
  return 1;
}

int Trova_i(int a)
{
  int b;

  for (b=0; b<3*n; b++)
    if ((A[b]!=4)&&(A5[b]==1)&&Free_Machine(a,b))
      return b;
  return 4*n;
}

int Verify2(int a) /*Verifies if "a" is an Unload operation*/
{
  int c;

  c=0;
  while(c<n){
    if(a==(2+3*c))
      return 1;
    else
      c++;
  }
  return 0;
}

int main()
{
  int instance;
  double Cmax_sum;

  instance=0;
  Cmax_sum=0;
  while (instance<10){
```

```
Instance_Generation();
t=0;
while (Finish()==0) {
  i=Find_i(t);
  while (i==(4*n)){
    t=t+1;
    for (j=0; j<3*n; j++)
      if ((A[j]==4)&&(t>=end[j])&&(Verify2(j)==0)
      &&(A5[j+1]==0)){
        A5[j+1]=1;
        A2ini[j+1]=t;
        }
    i=Find_i(t);
}
if (A[i]==2) {
  A3[ml]=t+A1[i];
  for (j=0; j<3*n; j++)
    if (A5[j]==1)
  A2[j]=t-A2ini[j];
  A[i]=4;
  end[i]=t+A1[i];
}
else {
  i=Find_max(t);
  while (i==(4*n)){
    t=t+1;
    for (j=0; j<3*n; j++)
      if ((A[j]==4)&&(t>=end[j])&&(Verify2(j)==0)
      &&(A5[j+1]==0)){
        A5[j+1]=1;
        A2ini[j+1]=t;
        }
    i=Find_max(t);
  }
if (A6[i]==1) {
  A3[ml]=t+A1[i]+A1[i+1];
  for (j=0; j<num_mac; j++)
    if(AssMacTask[i+1][j]==1)
      ml=j;
    A3[ml]=t+A1[i]+A1[i+1];
    for (j=0; j<3*n; j++)
```

```
       if (A5[j]==1)
       A2[j]=t-A2ini[j];
       A[i]=4;
       A[i+1]=4;
       end[i]=t+A1[i]+A1[i+1];
       end[i+1]=t+A1[i]+A1[i+1];
    }
    else
      if (A6[i]==0){
      A3[ml]=t+A1[i];
      for (j=0; j<3*n; j++)
        if (A5[j]==1)
      A2[j]=t-A2ini[j];
      A[i]=4;
      end[i]=t+A1[i];
      }
    }
   for (j=0; j<3*n; j++)
    if ((A[j]==4)&&(t>=end[j])&&(Verify2(j)==0)&&(A5[j+1]==0)){
      A5[j+1]=1;
      A2ini[j+1]=t;
    }
  }
  printf("Cmax: %i",t);
  Cmax_sum=Cmax_sum+t;
  instance++;
  }
  printf("Averge Cmax: %0.2f",Cmax_sum/10);
}
```

Appendix B

In what follows we show how the models described in Section 3 can be implemented in the mathematical language AMPL [70] and solved by means of a mathematical problem solver such as CPLEX [36].

AMPL Codes for Model (3.1)-(3.5) in Chapter 3

File Makespan.mod; it minimizes the makespan as reported in (3.1)-(3.5).

SET

```
set TASK;
set TIMESLOT;
set ARC;
```

#PARAMETERS

```
param head{ARC} in TASK ;
param tail{ARC} in TASK ;
```

#VARIABLES

```
var tau ;
var x{TASK, TIME} binary;
```

#MODEL

```
minimize z:
        tau;
```

subject to

(3.2) {i in TASK, t in TIMESLOT}: tau >= t*x[i][t] ;

(3.3) {i in TASK}: sum {t in TIMESLOT} x[i][t] = 1 ;

(3.4) {l in ARC, i in TIMESLOT}: x[head[l]][t] + x[tail[l]][t] <=1 ;

This is an example of a data file associated with the model file Makespan.mod. We have supposed an instance with 10 tasks and a time horizon of 3 time slots. The incompatibility graph is formed by 10 arcs whose heads and tails are defined above.

File Makespan.dat; it contains the data of the model Makespan.mod

SET

set TASK := 1 2 3 4 5 6 7 8 9 10;
set TIMESLOT:= 1 2 3 ;

#PARAMETERS

param: ARC: head tail :=

1 1 4
2 1 5
3 1 7
4 2 4
5 6 5
6 2 6
7 3 5
8 3 6
9 5 7
10 4 6;

AMPL Codes for Model (3.13)-(3.18) in Chapter 3

#File Deviation.mod; it minimizes the deviation as reported in (3.6) - (3.11).

SET

set TASK;
set TIMESLOT;
set ARC;
set KTYPE;

PARAMETERS

param head{ARC} in TASK ;
param tail{ARC} in TASK ;
param req{TASK,KTYPE} ;
param mu{KTYPE};

VARIABLES

var x{TASK, TIME} binary;
var dev{TIME};
var tc{KTYPE, TIME};

#MODEL

min z:
        sum{t in TIMESLOT} dev[t]

subject to
(3.7) {i in TASK}: sum {t in TIMESLOT} x[i][t] = 1 ;
(3.8) {t in TIMESLOT, k in KTYPE}: tc[k][t] = sum {i in TASK}
(req[i][k]*x[i][t]);
(3.9) {t in TIMESLOT}: dev[t]=(sum{k in KTYPE}
(abs(mu[k]-tc[k][t])/card[KTYPE]);
(3.9′) {k in KTYPE}: mu[k]=sum {i in TASK}
req[i][k]/card(TIMESLOT);
(3.10) {l in ARC, t in TIMESLOT}: x[head[l]][t] + x[tail[l]][t] <=1 ;

This is an example of a data file associated with the model file Deviation.mod.
We have supposed an instance with 4 tasks and a time horizon of 3 time slots.
The incompatibility graph is formed by 4 arcs whose heads and tails are de-

fined above.

The Deviation.dat; it contains the data of the model Deviation.mod

SET

set TASK := 1 2 3 4 ;
set TIMESLOT:= 1 2 3 ;

#PARAMETERS

param: EDGE: head tail :=

1 1 4
2 1 3
3 2 4
4 3 4 ;

AMPL Codes for Model (3.19)-(3.24) in Chapter 3

File Peak.mod; it minimizes the peak as reported in (3.12) - (3-17).

SET

set TASK;
set TIMESLOT;
set ARC;
set KTYPE;

PARAMETERS

param head{ARC} in TASK;
param tail{ARC} in TASK;
param req{TASK,KTYPE};

VARIABLES

var x{TASK, TIME} binary;
var tc{KTYPE, TIME};
var peak;

MODEL

min z:
peak

subject to
(3.13) {i in TASK}: sum{t in TIMESLOT} x[i][t] = 1;
(3.14) {t in TIMESLOT, k in KTYPE}: tc[k][t] = sum {i in TASK}
(req[i][k]*x[i][t]);
(3.15) {t in TIMESLOT, k in KTYPE}: peak >= tc[k][t];
(3.16) {l in ARC, t in TIMESLOT} x[head[l]][t] + x[tail[l]][t] <= 1;

Glossary

Basic graph terminology

- A *graph* $G = (V, E)$ is a collection of two sets, i.e., V and E, where V is a finite set of *vertices* (or *nodes*), and E is a set of couples of vertices called *edges*.
- A directed graph (*digraph*), is a graph with directions assigned to its edges, that is $D = (V, A)$, is a digraph where V is a finite set of vertices, and A is a set of *ordered* couples of vertices called *arcs*.
- Two vertices are *adjacent* if there is an edge joining them.
- Two vertices are *independent* if there is no edge joining them.
- A set of vertices adjacent to a given vertex is called the *neighborhood* of that vertex.
- A set of mutually adjacent vertices is called *clique*.
- A set of mutually *non* adjacent vertices is called *independent set*.
- In a graph $G = (V, E)$, the *degree* of a node v is the number of edges containing v. In a digraph $D = (V, A)$, the *outdegree* of a node v is the number of arcs of the form (v, u), while the *indegree* of a node v is the number of arcs of the form (u, v).
- The density of a graph δ is defined as the probability that an edge exists between to nodes. Ii this way a graph with $\delta = 0.1$ (sparse graph) has probability 0.1 that any couple of nodes is joined by an edge, while for a graph with $\delta = 0.9$ (dense graph) the probability that any couple of nodes is joined by an edge is 0.9.
- A *matching* M of a graph $G = (V, E)$ is a subset of edges with the property that no two edges of M so that no two edges of M have a node in common.
- A *coloring* C of a graph $G = (V, E)$ is a mapping of its vertices to a set of labels (colors) so that no vertices connected by an edge have different labels. Equivalently, a coloring is a partition of the nodes into independent set.

- A *minimum coloring* is a coloring that uses as colors as possible. Equivalently, a coloring is a partition of nodes into the minimum number of independent sets.
- The *clique number* of a graph $G = (V, E)$ is the size of the largest possible clique of G and is denoted as $\omega(G)$.
- The *chromatic number* of a graph $G = (V, E)$ is the number of colors used by a minimum coloring of G and is denoted by $\chi(G)$.
- A *path* is an alternating sequences of nodes and edges starting and ending with a node so that two subsequent edges have one node in common.
- A *cycle* is a path where the first and last node coincide.
- A Directed Acyclic Graph (DAG) is a directed graph without cycles.

Basic graph representations

- Let $G = (V, E)$ a graph whose vertices have been arbitrarily ordered $v_1, v_2, \ldots, v_n$. The *adjacency matrix* $A = a_{ij}$ of G is an $n * n$ matrix with entries

$$a_{ij} = 0 \text{ if } v_I, v_j \notin E,$$

$$a_{ij} = 1 \text{ if } v_I, v_j \in E$$

- If for each vertex v_i of G we create a list $Adj(v_i)$n containing those vertices adjacent to v_i we can represent the graph by the *adjacency list*.

References

[1] Agnetis, A., 2000, Scheduling no-wait robotic cells with two and three machines, *European Journal of Operational Research*, 123 (2), 303–314.

[2] Agnetis, A., Arbib, C., Lucertini, M., Nicolò, F., 1990, Part routing in Flexible Assembly Systems, *IEEE Transactions on Robotics and Automation*, 6 (6), 697–705.

[3] Agnetis, A., Lucertini, M., Nicolò, F., 1993, Flow Management in Flexible Manufacturing Cells with Pipeline Operations, *Management Science*, 39 (3), 294–306.

[4] Agnetis, A., Macchiaroli, R., 1998, Modelling and Optimization of the Assembly Process in a Flexible Cell for Aircraft Panel Manufacturing, *International Journal of Production Research*, 36 (3), 815–830.

[5] Agnetis, A., Pacciarelli, D., 2000, Part sequencing in three-machine no-wait robotic cells, *Operations Research Letters*, 27, 185–192.

[6] Akturk, M. S. E., 1999, An exact tool allocation approach for CNC machines, *Int. J. Computer Integrated Manufacturing*, 2, 129–140.

[7] Akturk, M. S. E., Avci, S., 1996, Tool allocation and machining conditions optimization for CNC machines, *European journal of Operational Research*, 94, 335–348.

[8] Akturk, M. S. E., Avci, S., 1996, Tool magazine arrangement and operations sequencing on cnc machines *Computers Operations research*, 23 (11), 1069–1081.

[9] Akturk, M. S. E., Onen, S., 1999, Joint lot sizing and tool management in a CNC enviroment, *Computers in Industry*, 40, 61–75.

[10] Alpert, C.J., Kahng, A.B., 1995, Recent directions in netlist partitioning: A survey, *VLSI Journal*, 19 (1-2), 1–81.

[11] Aneja, Y. P., Kamoun, H., 1999, Scheduling of parts and robot activities in a two machine robotic cell, *Computer and Operation Research*, 26 (4), 297–312.

[12] Asfahl, C. R., 1985, *Robots and Manufacturing Automation*, Wiley, New York, NY.

[13] Askin R. G., and Standridge C. R., 1993, *Modeling and analysis of Manufacturing Systems*, John Wiley and Soons,Inc., New York.

[14] Aspnes, J., Azar, Y., Fiat, A., Plotkin, S., Waarts, O., 1993, On-line machine scheduling with applications to load balancing and virtual circuit routing, in "Proc. 25th Annual ACM Symposium on Theory of Computing", 623–631.

[15] Awerbuch, B., Azar, Y., Plotkin, S., 1993, Throughput competitive on-line routing, in "Proc. 34th IEEE Annual Symposium on Foundations of Computer Science", 32–40.

[16] Awerbuch, B., Azar, Y., Plotkin, S., Waarts, O., 1994, Competitive routing of virtual circuits with unknown duration, in "Proc. 5th ACM SIAM Symposium on Discrete Algorithms", 321–327.

[17] Azar, Y., Broder, A., Karlin A., 1992, On-line load balancing, in "Proc. 33rd IEEE Annual Symposium on Foundations of Computer Science", 218–225.

[18] Azar, Y., Broder, A., Karlin A., 1994, On-line load balancing, *Theoretical Computer Science*, 130 (1), 73–84.

[19] Azar, Y., Epstein, L., 2004, On-line Load Balancing of Temporary Tasks on Identical Machines, *SIAM Journal on Discrete Mathematics*, 18 (2), 347–352.

[20] Azar, Y., Kalyanasundaram, B., Plotikin, S., Pruhs, K.R., Waarts, O., 1997, On-line load balancing of temporary tasks, *Journal of Algorithms*, 22 (1), 93–110.

[21] Azar, Y., Naor, J., Rom, R., 1992, The competitiveness of on-line assignment, in "Proc. 3rd ACM SIAM Symposium on Discrete Algorithms", 203–210.

[22] Baker, K. R., 1976, *Introduction to Sequencing and Scheduling*, John Wiley and Sons.

[23] Baker, A. D., Merchant, M. E., 1993, Automatic factories: How will they be controlled, *IEEE Potentials*, 12 (4), 15–20.

[24] Bakker, J.J., 1989, DFMS: Architecture and Implementation of a Distributed Control System for FMS, Delft, Techn. Univ., Diss.

[25] Bandelloni, M., Tucci, M., Rinaldi, R., 1994, Optimal resource levelling using non-serial dynamic programming, *European Jounal of Operational Research*, 78, 162–177.

[26] Baptiste, P., Le Pape, C., 2000, Constraint Propagation and Decomposition Techniques for Highly Disjunctive and Highly Cumulative Project Scheduling Problems, *Constraints*, 5, 119–139.

[27] Barash, Upton, Veeramani, 1992, Cutting tool management in computer integrated manufacturing, *The international journal of flexible manufacturing systems*, 3 (4), 237–265

[28] Bard J. F., 1998, A heuristic for minimizing the number of tool switches on a flexible machine, *IIE Transactions*, 20 (4).

[29] Baroum, S.M., Patterson, J.H., 1996, The Development of Cash Flow Weight Procedures for Maximizing the Net Present Value of a Project, *Journal of Operations Management*, 14, 209–227.

[30] Bartal, Y., Fiat, A., Karloff, H. Vohra R., 1992, New algorithms for an ancient scheduling problem, in "Proc. 24th Annual ACM Symposium on Theory of Computing".

[31] Bartusch, M., Möhring, R.H., Radermacher, F.J., 1988, Scheduling project networks with resource constraints and time windows, *Annals of Operations Research* 16, 201–240.

[32] Bauer, A., Bowden, R., Browne, J., Duggan, J., Lyons, G., 1991, *Shop Floor Control Systems, From design to implementation*, Chapman and Hall, London.

[33] Belady, L. A., 1966, A study of replacement algorithms for a virtual-storage computer. *IBM Systems Journal*, 5, 78–101.

[34] Bianco, L., Blazewicz, J., Dell'Olmo, P., Drozdowski, M., 1994, Scheduling Preemptive Multiprocessor Tasks on Dedicated Processors. *Performance Evaluation* 20 (4), 361–371.

[35] Bianco, L., Dell'Olmo, P., Speranza, M.G., 1994, Nonpreemptive scheduling of indipendent tasks with prespecified processors allocation, *Naval Research Logistics* 41, 959–971.

[36] Bixby, N., Boyd, E., 1996, Using the CPLEX callable library. CPLEX Optimization Inc., Houston, Texas.

[37] Blazewicz, J., Ecker, K., Pesch, E., Schmidt, G., Weglarz, J., 1996, *Scheduling Computer and Manufacturing Processes*, SpringerVerlag, Heidelberg.

[38] Blazewicz, J., Finke, G, 1994, Scheduling with resource management in manufacturing systems, *European Journal of Operational Research*, 76, 1–14.

[39] Blazewicz, J., Kubiak, W., Sethi, S. P., Sorger, G., Srikandarajah, C., 1992, Sequencing of parts and robot moves in a robotic cell, *International Journal of Flexible Manufacturing Systems*, 4, 331–358.

[40] Borodin, A., Linial, N., Saks, M., 1992, An optimal online algorithm for metrical task systems, *J. Assoc. Comput. Mach.*, 39, 745–763.

[41] Brauner N., Finke G., 2001, Optimal moves of the material handling system in a robotic cell, *International Journal of Production Economics*, 74, 269–277.

[42] Brinkmann, K., Neumann, K., 1996, Heuristic Procedures for Resource-Constrained Project Scheduling with Minimal and Maximal Time Lags: The Resource-Levelling and Minimum Project-Duration Problems, *Journal of Decision Systems* 5, 129–156.

[43] Brucker P., 2001, Scheduling Algorithms, Springer-Verlag.

[44] Brucker, P., Drexl, A., Mohring, R., Neumann, K., Pesch, E., 1999, Resource-constrained project scheduling: Notation, classification, models, and methods. *European Journal of Operational Research* 112, 3–41.

[45] Buffa, E. S., Miller, J. G., 1979, Production-Inventory Systems: Planning and Control, Homewood, Illinois.

[46] Bui, T., Jones, C., 1993, A heuristic for reducing fill in sparse matrix factorization, in "Proc. 6th SIAM Conf. Parallel Processing for Scientific Computing", 445–452.

[47] Caramia, M., Dell'Olmo, P., 2001, Iterative Coloring Extension of a Maximum Clique, *Naval Research Logistics* 48, 518–550.

[48] Caramia, M., Dell'Olmo, P., Italiano, G.F., 2001, New Algorithms for Examination Timetabling, Lecture Notes in Computer Science, 1982, 230–242.

[49] Caramia, M., Dell'Olmo, P., Onori, R., 2004, Minimum Makespan Task Sequencing with Multiple Shared Resources, *Robotics and Computer Integrated Manufacturing*, 20 (1).

[50] Castelfranchi, C., 1995, Guarantees for Autonomy in Cognitive Agent Architecture, in M. Wooldridge and N. Jennings (Eds.) *Intelligent Agents: Theories, Architectures, and Languages*, Lecture Notes in Artificial Intelligence, 890, Springer-Verlag, Heidelberg, 56–70.

[51] Chryssolouris, G., 1991, An Approach for Allocating Manufacturing Resources to Production Tasks, *Journal of Manufacturing Systems*, 10 (5), 368.

[52] Chryssolouris, G., 1992, *Manufacturing Systems, Theory and Practice*, Springer-Verlag, New York.

[53] Christensen J., 1994, Holonic Manufacturing Systems - Initial Architecture and standard directions, in Proceedings of the first European Conference on Holonic Manufacturing Systems, Hannover.

[54] Chu, C., Proth, J.M., 1996, Single machine scheduling with chain structures procedence constraints and separation time windows, *IEEE Transactions on Robotics and Automation*, 12 (6), 835–844.

[55] Ciobanu, G., 1972, A multi-project programming model with resource levelling. *Economic Computation and Economic Cybernetics Studies and Research* 3, pp. 61–68.

[56] Crama Y., 1997, Combinatorial opimization models for production scheduling in automated manufacturing systems, *European Journal of Operational Research* 99, 136–153.

[57] Crama Y., Klundert van de J., 1996, *The approximability of tool management problems* Research Memoranda of the METEOR, Maastricht Research School of Economics of Technology and Organization, 34.

[58] Crama Y., Kolen A. W. J., Oerlemans A. G., Spieksma F. C. R. , 1994, Minimizing the number of tool switches on a flexible machine. *The international journal of flexible manufacturing systems* 6, 33–54.

[59] Crama, Y., Van De Klundert, J., 1997, Cyclic scheduling of identical parts in a robotic cell, *Operations Research*, 45, 952–965.

[60] Crama, Y., Van De Klundert, J., 1997, Robotic flowshop scheduling is strongly NP-complete, Ten Years LNMB, W.K. Klein Haneveld, O.J. Vrieze and L.C.M. Kallenberg (eds.), CWI Tract 122, Amsterdam, The Netherlands, 277–286.

[61] Dell'Amico, M., Maffioli, F., and Martello, S. (eds.), 1997, Annotated Bibliographies in Combinatorial Optimization, Wiley.

[62] Demeulemeester, E.L., 1995, Minimizing resource availability costs in time-limited project networks, *Management Science* 41 (10), 1590–1598.

[63] Detand J., 1993, A Computer Aided Process Planning System Generating Non- Linear Process Plans, PhD Thesis, KU Leuven.

[64] Djellab H., Djellab K., Gourgand M., 2000, A new heuristic based on a hypergraph representation for the tool switching problem. *International Journal of Production Economics*, 64, 165–176.

[65] Duffie, N. A., 1990, Synthesis of heterarchical manufacturing systems, *Computer Science in Industry*, Special Issue: Josef Hatvany Memorial: Total Integration - Analysis and Synthesis, 14, 167–174.

[66] Duffie, N. A., Piper, R. S., Humphrey, B. J., Hartwick, J. P., 1998, Fault-tolerant heterarchical control of heterogeneous manufacturing system entities, *Journal of Manufacturing Systems*, 7 (4), 315–327.

[67] Duffie, N. Duffie, N. A., Prabhu, V. V., 1994, Real-time distributed scheduling of heterarchical manufacturing systems, *Journal of Manufacturing Systems*, 13 (2), 94–107.

[68] El Magrini, H., Teghem, J., 1996, Efficiency of metaheuristics to schedule general flexible job-shop, Pre-prints of the Ninth International Working Seminar on Production Economics, Innsbruck, Austria, 2, 343–363.

[69] Fiduccia, C.M., Mattheyses, R.M., 1982, A linear-time heuristic for improving network partitions, in "Proceedings of the 19th ACM/IEEE Design Automation Conference", 175–181.

[70] Fourer, R., Gay, D.M., Kernighan, B.W., 1993, AMPL- A modelling language for mathematical programming. Boyd & Fraser Publishing Company, Danvers, Massachussetts.

[71] French S., 1982, *Sequencing and Scheduling, An Introduction to the Mathematics of the Job-Shop*, John Wiley and Sons, Chichester, England.

[72] Garey, M.R., Johnson, D.S., Stockmeyer, L., 1976, Some simplified NP-complete graph problems, *Theoretical Computer Science*, 1, 237–267.

[73] Garey, M.R., Johnson, D.S., 1979, Computers and Intractability: A Guide to the Theory of $\mathcal{NP}$-completeness, W.H. Freeman & Company, San Francisco.

[74] Glass, C.A., Shafransky, Y.M., Strusevich, V.A., 2000, Scheduling for parallel dedicated machines with a single server, *Naval Research Logistics*, 47, 304–328.

[75] Glover, F., 2000, Computing Tools for Modelling Optimization and Simulation: Interfaces in Computer Science and Operations Research. M. Laguna and J.L. Gonzales-Valarde, eds., Kluwer Academic Publishers, 1–24.

[76] Goldberg, M.K., Burnestein, M., 1983, Heuristic improvement techniques for bisection of vlsi networks, in "Proc. IEEE Intl. Conf. Computer Design", 122–125.

[77] Goldratt E.M., Cox J., 1984, *The Goal, a Process of Ongoing Improvement, North River Press*, Croton-on-Hudson (N.Y.).

[78] Gol'stejn, E.G., Dempe S., 2002, A minimax resource allocation problem with variable resources, *European Journal of Operational Research*, (136) 1, 46–56.

[79] Gomez, Lorena, *Modelagem de sistemas de Manufatura flexiveis considerando restricones temporais e a capacidade do magazine* avaiable at: http://www.lac.inpe.br/ lorena/arthur/asmf2.PDF

[80] Graham, R. L., 1996, Bounds for certain multiprocessing anomalies, *Bell Syst. Tech. J.*, 45, 1563–1581.

[81] Graham, R.L., Lawler, E.L., Lenstra, J.K., Rinnooy Kan, A.H.G., 1979, Optimization and approximation in deterministic sequencing and scheduling: A survey, *Ann. Discrete Math.*, 5, 287–326.

[82] Gray, Seidmann, Stecke, 1993, A synthesis of decision models for tool management in automated manufacturing, *Management science*, 39 (5), 549–567.

[83] Gronalt, M., Grunow, M., Gunther, H.O., Zeller, R., 1997, A Heuristic for Component Switching on SMT Placement Machines, *International Journal of Production Economics*, 53, 181–190.

[84] Gunther, H.O., Gronalt, M., Zeller, R., 1998, Job Sequencing and Component Set-up on a Surface Mount Placement Machine, *Production planning & Control*, 9 (2), 201–211.

[85] Hall, N.G., Kamoum, H., Sriskandarajah C., 1997, Scheduling in robotic cells: classification, two and three machine cells, *Operations Research*, 45 (3), 421–439.

[86] Hall, N.G., Kamoun, H., Sriskandarajah C., 1998, Scheduling in robotic cells: complexity and steady state analysis, *European Journal of Operational Research*, 109, 43–65.

[87] Hall, N.G., Potts, C.N., Sriskandarajah, C., 2000, Parallel machine scheduling with a common server, *Discrete Applied Mathematics*, 102 (3), 223–243.

[88] Hall, N.G., Sriskandarajah, C., 1996, A survey on machine scheduling problems with blocking and no-wait in process, *Operations Research*, 44, 510–525.

[89] Hanen, C., Munier, A., 1995, A Study of the Cyclic Scheduling Problem on Parallel Processors, *Discrete Applied Mathematics* 57, 167-192,

[90] Hart, J.P., Shogan, A.W. (1987) Semi-greedy heuristics: An empirical study. *Operations Research Letters* 6, 107–114.

[91] Hartley, J., 1983, *Robots at Work*, North-Holland, Amsterdam.

[92] Hendrickson, B., Leland, R.W., 1993, A multilevel algorithm for partitioning graphs, tech. rep., Sandia National Laboratories, Alburquerque, NM.

[93] Hendrickson, B., Leland, R.W., Plimpton, S., 1995, An efficient parallel algorithm for matrix-vector multiplication, *Int. J. High Speed Computing*, 7 (1), 73–88.

[94] Hendrickson, B., Leland, R.W., 1995, The Chaco user's guide, version 2.0, tech. rep., Sandia National Laboratories, Alburquerque, NM.

[95] Herroelen, W., De Reyck, B., Demeulemmester, E., 1998, Resource-constrained project scheduling: a survey of recent developments, *Computers and Operations Research*, 25 (4), 279–302.

[96] Hoitomt, D., Luh, P., Max, E., Pattipati, K., 1989, Schedule Generation and Reconfiguration for Parallel Machines, in "Proc. of the IEEE int. Conf. on Robotics and Automation", 528–533.

[97] Hoitomt, D., Luh, P., Pattipati, K., 1990, A Lagrangian Relaxation Approach to Job Shop Scheduling Problems, in "Proc. IEEE Int. Conference on Robotics and Automation", Cincinnati, Ohio, USA, 1944–1949.

[98] Jeng, W.D., Lin, J.T., Wen, U.P., 1993, Algorithms for sequencing robot activities in a robt-centered parallel-processor workcell, *Computer and Operations Research*, 20 (2), 185–197.

[99] Karger, D., Phillips, S., Torng, E., 1993, A better algorithm for an ancient scheduling problem, unpublished manuscript.

[100] Karlin, A. R., Manasse, M. S., Rudolph, L., Sleator, D. D., 1998, Competitive snoopy caching, *Algorithmica*, 1 (3), 70–119.

[101] Karp, R., Vazirani, U., Vazirani, V., 1990, An optimal algorithm for on-line bipartite matching, in "Proc. 22nd Annual ACM Symposium on Theory of Computing", 352–358.

[102] Karypis, G., Kumar, V., Aggarwal, R., Shekhar, S., Hypergraph partitioning using multilevel approach: applications in VLSI domain, *IEEE Transactions on VLSI Systems*, to appear.

[103] Karypis, G., Kumar, V., Grama, A., Gupta, A., 1994, Introduction to Parallel Computing: Design and Analysis of Algorithms. Redwood City, CA: Benjamin/Cummings Publishing Company.

[104] Karypis, G., Kumar, V., 1998, MeTiS A Software Package for Partitioning Unstructured Graphs, Partitioning Meshes, and Computing Fill-Reducing Orderings of Sparse Matrices Version 3.0, University of Minnesota, Department of Comp. Sci. and Eng., Army HPC Research Center, Minneapolis.

[105] Karypis, G., Kumar, V., Aggarawal, R., Shekhar, S., 1998, hMeTiS A Hypergraph Partitioning Package Ver-sion 1.0.1. University of Minnesota, Department of Comp. Sci. and Eng., Army HPC Research Center, Minneapolis.

[106] Kats, V., Levit, V.E., Levner E., 1997, An improved algorithm for cyclic flowshop scheduling in a robotic cell, *European Journal of Operational Research*, 97, 500–508.

[107] Kernighan, B.W., Lin, S., 1970, An efficient heuristic procedure for partitioning graphs, *The Bell System Technical Journal*, 49, 291–307.

[108] Kernighan, B.W., Scheikert, D.G., 1972, A proper model for the partitioning of electrical circuits, in "Proceedings of the 9th ACM/IEEE Design Automation Conference", 57–62.

[109] Kise, H, 1991, On an automated two-machines flowshop scheduling problem with infinite buffer, *Journal of the Operations Research Society of Japan* 34 (3), 354–361.

[110] Kise, H., Shioyama, T., Ibaraki, T., 1991, Automated two machines flowshop scheduling: a solvable case, *IIE Transactions*, 23 (1), 10–16.

[111] Klein, R., 2000, Scheduling of Resource-Constrainted Projects. Kluwer.

[112] Klein, R., Scholl, A., 1996, Maximizing the production rate in simple assembly line balacing - A branch and bound procedure, *European Journal of Operational Research*, 91, 367–385.

[113] Koestler, A., 1967, *The Ghost in the Machine*, Hutchinson and Co, London.

[114] Kouvelis, P., 1991, An optimal tool selection procedure for the initial design phase of a flexible manufacturing system, *European Journal of Operational Research*, 55, 201–210.

[115] Lam, W.T., Wong, P., Ting, H.F., To, K.K., 2002, On-line Load Balancing of Temporary Tasks Revised, *Theoretical Computer Science*, 270 (1-2), 325–340.

[116] Lawler, E.L, Lenstra, J.K., Rinnooy Kan, A., Shmoys, D.B., 1993, Sequencing and Scheduling: Algorithms and Complexity, in S.C. Graves, A.H.G. Rinnooy Kan, P.H. Zipkin (eds.) Logisitcs of production and inventory, 445–522.

[117] Lei, M., Mitchell, M.T., 1997, A preliminary Research of Market Appraoch for Mass Customization Assembly Planning and Scheduling, Technical Report Hong Kong University of Science and Technology, Dept. of Industrial Engineering and Engineering Management.

[118] Leitao, P., Restivo, F., 1999, A Layered Approach to Distributed Manufacturing, in proceedings of 1999 Advanced Summer Institute, Belgium.

[119] Lengauer, T., 1990, *Combinatorial Algorithms for Integrated Circuit Layout*, John Wiley and Sons.

[120] Leondes, C., 2001, *Computer-Integrated Manufacturing*, CRC Press.

[121] Leung, L. C., Maheshwari, S. K., Miller, W. A., 1996, Concurrent part assignment and tool allocation in FMS with material handling considerations *European journal of operational research*, 94, 335–348.

[122] Levner, E., Kats, V., Levit, V.E., 1997, An improved algorithm for cyclic flowshop scheduling in a robotic cell, *European Journal of Operational Research*, 97, 500–508.

[123] Lofgren, C.B., McGinnis, L.F., 1986, Dynamic Scheduling for Flexible FCC Assembly, *IEEE ICSMC* 2, 1294–1297.

[124] Logendran, R., Sriskandarajah, C., 1996, Sequencing of robot and parts in two machine robit cells, *International Journal of Production Research*, 34, 3447–3463.

[125] Luh, P., Hoitomt, D., 1993, Scheduling of Manufacturing Systems Using the Lagrangian Relaxation Technique, *IEEE Trans. on Automatic Control*, 38 (7).

[126] Maimon, O.Z., Braha, D., 1998, A Genetic Algorithm Approach to Scheduling PCBs on a Single Machine, *International Journal of Production Research*, 36 (3), 761–784.

[127] Makarov, I.M., Rivin, E.I., 1990, *Modeling of Robotic and Flexible Manufacturing Systems*, Hemisphere Pub.

[128] Malone, T., Crowston, K., 1994, The Interdisciplinary Study of Coordination, *ACM Computing Surveys*, 26 (1), 87–119.

[129] Manasse,M. S., McGeoch, L. A., Sleator, D. D., 1988, Competitive algorithms for online problems, in "Proc. 20th Annual ACM Symposium on Theory of Computing", 322–332.

[130] Matzliach, B., Tzur, M., 1998, The online tool switching problem with non uniform tool size, *Int. Journal of production research*, 36 (12), 3407–3420.

[131] Matzliach, B., Tzur, M., 2000, Storage Management of Items in Two Levels of Availability, *European Journal of Operational Research*, 121, 363379

[132] MESA, 1997, Manufacturing Execution Systems, available at: http://www.mesa.org/.

[133] Mingozzi, A., Maniezzo, V., Ricciardelli, S., Bianco, L., 1994, An exact algorithm for project scheduling with resource constraints based on a new mathematical formulation, *Management Science*, 44, 714–729.

[134] Moder, J.J., Phillips, C.R., 1970, Project Management with CPM and PERT, Van Nostrand Reinhold, New York.

[135] Moder, J.J., Phillips, C.R., Davis, E.W., 1983, Project Management with CPM and PERT and Project Diagramming, Van Nostrand Reinhold, New York.

[136] Mokotoff, E., 1999, Scheduling to Minimize the Makespan on Identical Parallel Machines: An LP-Based Algorithm, *Investigacion Operativa*, 8, 97–108.

[137] Mokotoff, E., 2002, An exact algorithm for the Identical Parallel Machine Scheduling Problem, *European Journal of Operational Research*, to appear.

[138] Neumann K., Zimmermann J., 1999, Resource levelling for projects with schedule-dependent time windows, *European Journal of Operational Research*, 117 (3), 591–605.

[139] Neumann K., Zimmermann J., 2000, Procedures for resource leveling and net present value problems in project scheduling with general temporal and resource constraints, *European Journal Of Operational Research*, 127 (2), 425-443.

[140] Nwana, H., Lee, L., Jennings, N., 1996, Coordination in Software Agent Systems, *BT Technology Journal*, 14 (4), 79–88.

[141] Okino, N., 1993, Bionic Manufacturing System, in J. Peklenik (Ed.) CIRP, *Flexible Manufacturing Systems: Past-Present-Future*, 73–95.

[142] Papadimitriou C. H., Steigliz K., 1982, Combinatorial Optimization: Algorithms and Complexity, Prentice-Hall.

[143] Phan Huy, T., 2000, Constraint Propagation in Flexible Manufactoring, Springer, Berlin.

[144] Phillips S., Westbrook, J., 1993, Online load balancing and network flow, in "Proc. 25th Annual ACM Symposium on Theory of Computing", 402–411.

[145] Pinedo, M., 1994, *Scheduling: theory, algorithms and systems*, Prentice Hall.

[146] Privault, C., Finke, G., 1995, Modeling a tool switching problem on a single NC machine, *Journal of Intelligent Manufacturing*, 6, 87–94.

[147] Qu, C.W., Ranka, S., 1997, Parallel incremental graph partitioning, *IEEE Trans. Parallel and Distributed Systems*, 8 (8), 884–896.

[148] Rajkumar, K., Narendran, T.T., 1998, A Heuristic for Sequencing PCB Assembly to Minimize Set-up Times, *Production Planning & Control*, 9 (5), 465–476.

[149] Resende, M.G.C., Riberio, C.C., 2002, State of the Art Handbook in Metaheuristics, F. Glover and G. Kochenberger, eds., Kluwer Academic Publisher.

[150] Rembold, U., Nnaji, B.O. , Storr, A., 1993, *Computer Integrated Manufacturing and Engineering*, Addison-Wesley Publishing Company, Wokingham, England.

[151] Rupe, J., Kuo, W., 1997, Solutions to a Modified Tool Loading Problem for a Single FMM, *International Journal of Production Research*, 35 (8), 2253-2268.

[152] Sadiq, M., Landers, T.L., Taylor, G.D., 1993, A Heuristic Algorithm for Minimizing Total Production Time or A Sequence of Jobs on A Surface Mount Placement Machine, *International Journal of Production Research*, 31 (6), 1327–1341.

[153] Schaffter, M.W., 1997, Scheduling with forbidden sets, *Discrete Applied Mathematics*, 72, 155–166.

[154] Seidel, D., Mey, M., 1994, *IMS - Holonic Manufacturing Systems: System Components of Autonomous Modules and their Distributed Control*, HMS project (IMS-TC5).

[155] Serafini, P., Ukowich, W., 1989, A mathematical model for periodic scheduling problems, *SIAM Journal on Discrete Mathematics*, 2, 550–581.

[156] Sethi, S. P., Sriskandarajah, C., Sorger, G., Blazewicz, J., Kubiak, W., 1992, Sequencing of Parts and Robot Moves in a Robotic Cell, *Int. J. of Flexible Manufacturing Systems*, 4, 331–358.

[157] Shmoys, D., Wein, J., Williamson, D. P., 1991, Scheduling parallel machines on-line, in "Proc. 32nd IEEE Annual Symposium on Foundations of Computer Science", 131–140.

[158] Sihn, W., 1997, The Fractal Factory: A Practical Approach to Agility in Manufacturing, in "Proceedings of the 2nd World Congress on Intelligent Manufacturing Processes and Systems", 617–621.

[159] Sleator, D. D. Tarjan, R. E., 1985, Amortized efficiency of list update and paging rules, *Comm. Assoc. Comput. Mach.*, 28 (2), 202–208.

[160] Sousa, P., Ramos, C., 1999, A Distributed Architecture and Negotiation Protocol for Scheduling in Manufacturing Systems, *Computers in Industry*, 38 (2), 103–113.

[161] Sousa, P., Silva, N., Heikkila, T., Kollingbaum, M., Valckenaers, P., 1999, Aspects of cooperation in Distributed Manufacturing Systems, in "Proceedings of the 2nd International Workshop on Intelligent Manufacturing Systems (IMS-Europe'99)", Leuven, Belgium, 695–717.

[162] Syacara, K., 1989, Multi-Agent Compromise via Negotiation, in L. Gasser and M. Huhns (Eds.) *Distributed Artificial Intelligence 2*, Morgan Kaufmann.

[163] Stecke, K.E., 1983, Formulation and solution of nonlinear integer production planning problems for flexible manufacturing systems. *Management Science*, 29 (3), 273-288.

[164] Tang, C. S., Denardo, E. V., 1998, Model arising from a flexible manufacturing machine, part I: minimization of the number of tool switches, *Operation Research*, 36 (5), 767–777.

[165] Tang, C.S., Denardo, E.V, 1988, Models arising froma flexible manufacturing machine, part II: minimization of the number of switching instants, *Operations Research*, 36(5), 778-784.

[166] Tharumarajah, A., J. Wells, L. Nemes, 1996, Comparison of bionic, fractal and holonic manufacturing system concepts, *International Journal of Computer Integrated Manufacturing*, 9 (3), 217–226.

[167] Universiteit Maastricht web site   *Test Intsances*   available at: http://www.unimaas.nl/.

[168] Valckenaers, P., 1993, Flexibility for Integrated Production Automation, PhD. thesis K.U.Leuven.

[169] Valckenaers, P., Van Brussel, H., Bonneville, F., Bongaerts, L. and Wyns, J., 1994, IMS Test Case 5: Holonic Manufacturing Systems, Proceedings of the IMS Workshop at IFAC'94, Vienna.

[170] Van de Klundert, J., 1996, Scheduling problem in automated manufacturing, Dissertation no. 96-35, Faculty of Economics and Business Administration, University of Limburg, Maastricht.

[171] Vizvari, B., Demir, R., 1995, A Column Generation Algorithm to Schedule Identical Parallel Machines, *Pure Mathematicis and Applications* (PU.M.A.), 6, 287–299 (previously appeared as Rutcor Research Report 11-94).

[172] Younis M.A., Saad B., 1996, Optimal resource leveling of multi-resource projects, *Industrial Engineering*, 31 (1-2), 1–4.

[173] Warneke, H.J., 1993, *The Fractal Company*, Springer-Verlag.

[174] Wiest, J.D., Levy, F.K., 1969, A Management Guide to PERT/CPM, Prentice-Hall, Englewood Cliffs.

[175] Wilhelm, W.E., 1987, Complexity of sequencing tasks in assembly cells attended by one or two robots, *Naval Research Logistics*, 34, 3447–3463.

[176] Williamson, D.T.N., 1967, System 24 - A New Concept of Manufacture, Proceedings of the 8th International Machine Tool and Design Conference, 327–376, Pergamon Press.

Index

Printing: Krips bv, Meppel
Binding: Stürtz, Würzburg